# GLOBAL SUPPLY MANAGEMENT

A Guide to International Purchasing

# THE NAPM PROFESSIONAL DEVELOPMENT SERIES

**Michiel R. Leenders**
*Series Editor*

## VALUE-DRIVEN PURCHASING
**Managing the Key Steps in the Acquisition Process**
Michiel R. Leenders and Anna E. Flynn

## MANAGING PURCHASING
**Making the Supply Team Work**
Kenneth H. Killen and John W. Kamauff

## VALUE-FOCUSED SUPPLY MANAGEMENT
**Getting the Most Out of the Supply Function**
Alan R. Raedels

## PURCHASING FOR BOTTOM LINE IMPACT
**Improving the Organization through Strategic Procurement**
Lisa M. Ellram and Laura M. Birou

## GLOBAL SUPPLY MANAGEMENT
**A Guide to International Purchasing**
Dick Locke

# GLOBAL SUPPLY MANAGEMENT
## A Guide to International Purchasing

*The NAPM Professional Development Series*

*Dick Locke*

**Tempe, Arizona**

**McGraw-Hill**

New York   San Francisco   Washington, D.C.   Auckland   Bogotá
Caracas   Lisbon   London   Madrid   Mexico City   Milan
Montreal   New Delhi   San Juan   Singapore
Sydney   Tokyo   Toronto

# McGraw-Hill

A Division of The **McGraw·Hill** Companies

**Library of Congress Cataloging-in-Publication Data**

Locke, Dick.
    Global supply management / Dick Locke.
      p.   cm.
    Includes index.
    ISBN 0–7863–0797–8
    1. Industrial procurement—Management.  2. Foreign exchange.
  I. Title.
  HD39.5.L63   1996
  658.7'2—dc20                         95–50816

*Printed in the United States of America*
  2 3 4 5 6 7 8 9 0 BS 3 2 1 0 9 8 7

# *Preface*

This book grew out of my experiences at Hewlett-Packard Company (HP), where I managed international procurement offices (IPOs) in Asia, Europe, and North America.

In the early 1980s, I was managing a semiconductor procurement team at corporate headquarters. We bought from Japanese and European suppliers, but mainly through their sales subsidiaries in the United States. In some cases, we even dealt with independent representatives of those subsidiaries. I remember thinking that there must be a more effective way to deal with those suppliers.

I got my chance to prove this when I was asked to move to Tokyo and start HP's first four IPOs in Asia. From the time I arrived in mid-1984—starting with a desk, a phone, and no employees—until I left HP in 1993 to start my own consulting business, HP's IPOs grew from zero to several hundred million dollars in purchasing volume.

Those years were an intense learning experience. At the time, there were no comprehensive textbooks in the field of international purchasing. We learned as we went along. Based on results, we seemed to do most things right.

This book captures most of the lessons learned and adapts some of them for use in smaller companies. It is intended for two groups of users. The first group is those in manufacturing and retail companies who select and manage suppliers and who have relatively little experience in buying directly from foreign suppliers. The second group is staff and management of companies who are currently buying from foreign suppliers but want to check and update their skills against leading-edge techniques. Throughout the book, I have provided information that both large and small companies can use.

The book is organized into two parts. The first part covers key skills that are necessary in approaching foreign suppliers directly. I stress the advantages of dealing directly with foreign suppliers for several reasons, with the most important being cost. Dealing through foreign suppliers' representatives and subsidiaries in the United States is relatively easy but expensive. I have seen these organizations add markups as high as 74 percent to the cost of goods. I have not seen markups lower than 5 percent, even in arrangements for annual purchases of tens of millions of dollars. Five percent does not sound high, but on large volume it can be a multi-million-dollar markup.

The first necessary skill is an understanding of cultural differences and how they affect purchasing. This is a pervasive issue throughout international purchasing. Inexperienced buyers think it is important, but often mistake the surface aspects of a culture (such as how Japanese handle business cards) for the important part. The book explains the more fundamental differences, such as patterns of communication.

After covering culture, the book goes on to address skillful use of language, including body language. It then covers some important legal aspects of doing business overseas. This section includes information on some unique U.S. legal practices that do not work well outside of United States.

The next topic, and one that the book covers very thoroughly, is currency. This is an area in which buyers in other, smaller countries have an advantage over American buyers. In those countries, businesspeople routinely buy and sell in foreign currencies. Readers of this book will be able to get lower prices through skillful selection of currencies, and they will be able to maintain those prices through selection of the proper hedging techniques.

The chapters on currency contain some exercises to help readers check their understanding. I put special emphasis on this section because I believe that many companies' finance staffs do not understand international currency issues. Procurement departments will often need to lead the finance staff to a better understanding. If the finance staff does not become skillful with currencies, purchasing will be faced with high material costs.

The final skills covered are customs and logistics. This book is not intended to create logistics and customs experts. That task would require a much longer book. The chapters on these topics cover logistics and customs for buyers. The chapters will give buyers the ability to communicate with experts in the fields. They will enable readers to recognize opportunities to minimize duties and logistics costs. The chapter also covers some common problem areas, some of which I have learned about the hard way.

After covering the key skills, the book shows how to utilize those skills in a "procurement cycle" of deciding what to source internationally, searching for suppliers, selecting suppliers, negotiating a deal, and managing remote suppliers. In this section, the book keeps its focus on the differences between domestic and international purchasing.

Finally, there are three appendixes. The first contains answers to the tests of understanding that appear in the chapters on currency. The second is a sample supplier survey that many companies will find useful. The third appendix lists important facts about 14 countries that are the major importers to the United States. This appendix contains valuable information on the countries' exports to the United States, their culture compared with that of the United States, and the stability of their currency. This appendix also includes valuable phone and fax numbers of helpful organizations.

I hope that readers will find the book as enlightening as I found the experiences that enabled me to write it.

## STYLE NOTES

I wrote this book from a United States perspective. I will frequently be using the words "United States" and "dollars" in cases where I could have said "buyer's home country" or "buyer's currency." This is simply to avoid awkward sentence construction. The content is usually suitable for buyers in any country. Similarly, and with apologies to Canadian and Latin American readers, I will write "American" to mean "a person from the United States of America." This also simplifies sentence construction.

## ACKNOWLEDGMENTS

I want to thank Jeff Cooke of Hewlett-Packard for his helpful comments on the culture of some countries; Leslie Murphy of Summit Information Services for providing trade data; and Bob Cashill, former U.S. editor of *Electronic Components* magazine, for helping me with my writing style.

**Dick Locke**

# Contents

*Preface*                                                                      *v*

*Chapter One*
IT'S A DIFFERENT WORLD                                                          1

    First Problem: Locating Potential Suppliers   1
    Next Problem: Getting Quotes   3
    Visiting the Suppliers   4
    Doing Business   5
    Not Hypothetical   6
    Key Points   6

*Chapter Two*
CULTURAL DIFFERENCES: INTRODUCTION                                             7

    What Is Culture?   7
    Some Words of Caution   9
    Importance of Cultural Skills   9
    Getting to the Right People   11
    Americans' Reputation   11
    Key Points   11
    Resources and References   12

*Chapter Three*
CULTURAL VALUES                                                               13

    Culture's Consequences   13
       ***Feature: Cultural Miscues   16***
    Other Value Differences   21
    Key Points   22
    Resources and References   23

*Chapter Four*
COMMUNICATION DIFFERENCES                                    25

    Message Speed   25
        *Feature: Communication with Germans   26*
    High- or Low-Context   26
    Getting to "No"   29
    Key Points   30
    Resources and References   30

*Chapter Five*
OTHER BEHAVIORAL DIFFERENCES                                 31

    Space, Territory, and Privacy   31
    Privacy and Territory   32
    Time   32
        *Feature: Cultures and Time   34*
    Cultural Aspects of Time   35
    Appointments   36
    Ease of Completing Action Chains   36
    Gift Giving   36
    Key Points   37
    Resources and References   38

*Chapter Six*
OVERCOMING CULTURAL DIFFERENCES                              39

    The Six Key Differences   39
    Know Yourself and Your Culture   40
    Stereotyping Americans   41
    Hosting and Being Hosted   42
    Cultural Advice   42
    Final Advice   43
    Key Points   43
    Resources and References   44

*Chapter Seven*
# LANGUAGE                                                                           45

Communication Is Your Responsibility   45
Adjust Your Speaking Style   46
Using Interpreters   48
   *Feature: Where Did They Learn to Speak English, Anyway?*   *49*
Check for Understanding   50
Does the Supplier Have to Know English?   50
Key Points   51
Resources and References   51

*Chapter Eight*
# LAW                                                                                     53

U.S. Legal Practices   53
Other Legal Systems   53
Contracts   54
   *Feature: Sample Arbitration Clause*   *59*
CISG   60
   *Feature: CISG Signers*   *60*
Legal Priorities   61
Intellectual Property   61
Foreign Corrupt Practices Act   63
Reciprocity   63
Unions   64
Key Points   64
Resources and References   65

*Chapter Nine*
# INTRODUCTION TO CURRENCY                                       67

The Currency Problem   67
Converting Prices   69
Weaker and Stronger   70
Exchange Risk   71
Why Take the Risk?   73
   *Feature: Test of Understanding*   *74*
Key Points   74
Resources and References   75

*Chapter Ten*
CURRENCY SELECTION                                                          77

    Key Definitions   77
    Type of Product Purchased   81
    Manufacturer's Country   82
    Summary   83
    Key Points   84

*Chapter Eleven*
INTRODUCTION TO HEDGING                                                     85

    Spot Rates   86
    How Much Can Exchange Rates Change?   88
    Risk Management   90
        *Feature: Test of Understanding   91*
    Key Points   91

*Chapter Twelve*
HEDGING FOR FIXED DOLLAR COSTS                                              93

    Forward Contracts   93
        *Feature: Forwards as a Predictor   94*
    Cash Flow   95
    Effect of Forwards   95
    Futures Contracts   97
        *Feature: Test of Understanding   98*
    Key Points   99

*Chapter Thirteen*
CURRENCY OPTIONS                                                           101

    Options Defined   102
    Option Markets   102
    Option Example   102
    Options Compared to Spot and Forwards   104
    Controlling Option Costs   104
    Internal Hedging   106
    Cash Flow   106
        *Feature: Test of Understanding   106*
    Key Points   106

*Chapter Fourteen*
## HEDGING PRACTICES                                                                                  109

Behind the Scenes   109
Timing of Hedging   112
Forecast Risk   113
Timing Risk   115
Best Hedging Practice   115
Why Not Pay Dollars?   117
Multiple Currencies   118
 *Feature: Test of Understanding   119*
Key Points   119

*Chapter Fifteen*
## NONFINANCIAL PROTECTION                                                            121

Escape Clauses   121
Risk Sharing   122
 *Feature: Where Did that Half Percent Go?   122*
 *Feature: Test of Understanding   125*
Key Points   125

*Chapter Sixteen*
## LOGISTICS                                                                                               127

Who Should Handle the Logistics?   127
Inbound Freight Flow   127
Support Organizations   128
Air Freight   129
Ocean Freight   131
International Logistics and JIT   134
Key Points   134
Resources and References   135

*Chapter Seventeen*
## CUSTOMS                                                                                                  137

Customs Basics   137
Determining Duties   137
Special Trade Preferences   140
Clearing Customs   143

Key Points   146
Resources and References   146

*Chapter Eighteen*
CUSTOMS AND LOGISTICS PRACTICES                                    147

Reducing Duties   147
   *Feature: Red Ink   148*
Avoiding Problems   149
   *Feature: Innocents Abroad   151–52*
Focus Your Imports   153
Incoterms   153
Insurance   156
Key Points   157
Resources and References   158

*Chapter Nineteen*
PAYING THE SUPPLIER                                                159

Letters of Credit   159
Documents Against Payment   162
Wire Transfer   162
Obtaining Credit Terms   162
Key Points   164
Resources and References   164

*Chapter Twenty*
INTERNATIONAL PROCUREMENT OFFICES                                  165

Services   165
   *Feature: Global Price Management   167*
Costs   169
Staffing   169
IPO Locations   170
IPO Weaknesses   170
Should Your Company Open IPOs?   171
Key Points   172

*Chapter Twenty-One*
## CHANNEL MANAGEMENT                                        173

Seller's Subsidiary   173
Buying Directly   176
International Procurement Offices   176
Independent Representatives   177
*Feature: Collisions in the Channel   178*
Brokers   179
Taking Control of the Channel   179
Key Points   181

*Chapter Twenty-Two*
## SUPPLIER SELECTION: 1                                      183

Timing   183
Search Plan   183
*Feature: Where in the World?   184*
Country Information   185
*Feature: Country Development Examples   185*
National Trade Data Bank   186
*Feature: Political Stability   187*
Potential Suppliers   187
Company Information Sources   188
Initial Contact   190
Eliminating Suppliers   191
Price Quotations   191
Financial Data   192
Reference Checking   192
Supplier Visits   193
Surveys   193
*Feature: Financial Analysis   195*
Next Steps   200
Key Points   200
Resources and References   201

*Chapter Twenty-Three*
## SUPPLIER SELECTION: 2                                                        203

Landed Cost   203
Risk Analysis   207
Eliminating Suppliers   209
Final Negotiation   209
Final Decision   211
Contract Signing   211
Key Points   211

*Chapter Twenty-Four*
## MANAGING REMOTE SUPPLIERS                                                    213

Key Supplier-Management Skills   213
Global Supplier Management   216
Evaluate Sensibly   217
Maintain Pressure on Costs   217
Informal Communications   217
Formal Communication   219
Maintain Continuity   219
Ethics   220
Key Points   220

*Chapter Twenty-Five*
## CONCLUSION                                                                   221

Summary   221
Getting Started   221
And Finally,   222

*Appendix A*
## ANSWERS TO TESTS OF UNDERSTANDING                                           223

Chapter 9   223
Chapter 11   224
Chapter 12   224
Chapter 13   224
Chapter 14   225
Chapter 15   226

*Appendix B*
SUPPLIER SURVEY: FROM THE TEXT
OF GLOBAL SUPPLY MANAGEMENT                                        227

  Part One: General Information   227
  Part Two: Manufacturing   231
  Part Three: Quality   234
  Part Four: Materials   240

*Appendix C*
BUYER'S GUIDE TO KEY COUNTRIES                                     243

  Major Exports to the United States, 1994   243
  Trade Representation in the United States   243
  Useful Contacts in the Supplier's Country   243
  Buyer's Cultural Radar Diagram   244
  Monochronism   246
  Currency History   247
  Purchaser's Guide to Canada   248
  Purchaser's Guide to China   251
  Purchaser's Guide to France   254
  Purchaser's Guide to Germany   257
  Purchaser's Guide to India   260
  Purchaser's Guide to Italy   263
  Purchaser's Guide to Japan   266
  Purchaser's Guide to Malaysia   269
  Purchaser's Guide to Mexico   272
  Purchaser's Guide to Singapore   275
  Purchaser's Guide to South Korea   277
  Purchaser's Guide to Taiwan   280
  Purchaser's Guide to Thailand   283
  Purchaser's Guide to the United Kingdom   286

*Index*                                                                          *289*

# Chapter One

# It's a Different World

This book is directed toward two groups. The first group is made up of those who are starting to become involved in international purchasing. The second group is composed of those who are looking for a more cost-effective way to purchase than from local representatives or foreign suppliers' subsidiaries in the United States. The book focuses on the differences between international and domestic purchasing.

To understand what the differences are, let's look at what a hypothetical buyer might go through as he or she starts to get involved in international purchasing. The buyer is working for a medium-sized company whose only international buying has been a few standard products bought from U.S.-based subsidiaries of Japanese companies. The company is under intense cost pressure from some foreign competitors who are selling aggressively in the United States, and there are rumors that the company's domestic competitors are out "shopping the world" for lower-cost purchased products. While the new buyer is going to do a lot of things right, he's also going to make several mistakes. These mistakes will add 15 to 30 percent to his costs and will negate a lot of the benefits of shopping worldwide.

## FIRST PROBLEM: LOCATING POTENTIAL SUPPLIERS

In domestic purchasing, a buyer interested in locating new sources has a tried-and-true way of doing it. He or she picks up the local Yellow Pages, or a Thomas Register, or a trade magazine from the appropriate industry, and reviews advertisements. The buyer might review files of old business cards from salespeople who have called in the past and contact them.

How does the buyer locate the best sources internationally? It's a very big world to look in.

## Where in the World?

The buyer realizes that before looking for specific suppliers, he has to narrow down the list of countries that he will look in. This is a good move. He decides to approach this task on a macroeconomic level. First, he looks at the economics of manufacturing the product he is buying, which is an assembly of plastic molded parts. He discovers that typical manufacturing costs are moderate in labor content and heavy in petrochemical-based plastic material. There is also tooling needed for the plastic parts. The tooling will cost somewhere in the six-figure range.

From this information, the buyer decides to look for countries where labor costs are low and where there are good petrochemical resources. Some library research leads him to Mexico, Indonesia, and Nigeria as potentially qualified countries.

He doesn't know it yet, but he has made his first mistake. Macroeconomic considerations do not explain why a country becomes good at building things. A brief look at the Japanese car industry would demonstrate this. Japan is a high-cost, low-resource country that seems unlikely to be successful in car building. The latter chapters of this book give better ways of looking for appropriate countries.

However, the buyer has now decided to look at Indonesia, Mexico, and Nigeria. He calls the embassies of those three countries in Washington, D.C., and asks to speak to an export development officer. These officers immediately offer to help and quickly mail him lists of potential suppliers. These lists are basically company names, addresses, and fax and telephone numbers.

## Whom to Contact?

The buyer looks at the lists for a while, likes the sound of some of the company names, and sends faxes to about 10 suppliers in each country. In the fax he gives his company name and a rough description of the parts that he is interested in buying. He asks for some basic information on the potential seller's company. About half of the companies do not respond. English-speaking local representatives of about one-third of suppliers contact him within a week and announce that they are the U.S. reps of the suppliers. With a sense of relief, the buyer invites them to visit. He has just made the most common mistake buyers make. He has allowed a third party into the purchasing channel without establishing the third party's fees and services.

Many buyers will agree to dealing with local representatives (or with the supplier's sales subsidiaries) because they are concerned with the difficulties of dealing with remote suppliers. The buyers are concerned about language, culture, and communication difficulties. The first part of this book describes the skills required to be able to deal directly and effectively with foreign suppliers.

## NEXT PROBLEM: GETTING QUOTES

The buyer eventually gets to the point where he is comfortable asking suppliers to quote on his particular parts. He asks an assistant to put together the usual quote package.

### *Legal Problems*

The quote package contains drawings, forecasts of volumes, and the normal U.S. contract that the company asks suppliers to sign. The contract consists of 28 clauses spread over three pages, all written by the buyer's attorney. The attorney is quite proud of it and feels that it does a complete job of tying the supplier down. In fact, very few U.S. suppliers have signed the contract as it stands.

The buyer is not aware that the contract is of no practical use internationally. It will even offend many suppliers who are not used to some unusual features of American law. Two suppliers that would have been really good drop out. Chapter 8 of this book gives some ways to modify standard U.S. legal practice for international use.

### *Currency*

Several of the suppliers ask whether they should quote in dollars or in their own currency. The buyer considers briefly and discusses the issue with his company's treasurer. They decide the safe thing to do is to get quotes in dollars. They have just made the second biggest mistake buyers make. The seller is now faced with currency risk, so he raises his price approximately 5 percent.

Sadly, the buyer has only received an illusion of security for his extra 5 percent. Chapters 9 through 15 explain a better way to deal with foreign-currency prices.

## *Terms of Sale*

More suppliers ask what terms of sale the buyer wants. They state they would prefer to quote "Ex works" or "Free carrier." The buyer is unsure what these terms mean. After some research, he realizes that the prices will not include freight and duties. He's uncertain what the duty rates will be, and he has no ocean freight experience. Therefore, he asks the sellers to provide quotes for the parts delivered, with all duties paid, at a U.S. port of entry.

The remaining suppliers who were not selling through representatives drop out, stating they cannot deal with U.S. customs practices. Now the buyer has limited his potential supply base even further. He has also given up some significant opportunities for reducing duties. Chapters 17 through 19 of this book explain the meaning of international terms of sale, the advantages of a buying company's importing goods themselves, and the best ways to ask for quotations.

## VISITING THE SUPPLIERS

The buyer was disturbed at the number of suppliers who declined to quote. He also realized that some of the purported suppliers he was talking to were actually distributors and brokers. He had to repeat all the above steps to get a sufficient number of potential suppliers. In the process, he decided that Nigerian quotes were too high, and he discovered that good prices were available from Taiwan. At the end of the process, he had some quotes from suppliers in Taiwan and Indonesia that looked like they could save him 10 to 15 percent. While this is not as good as the savings he could have achieved, it was enough to justify looking further.

The buyer realized that the products he was having quoted were very strategic to his company's production and that he would have to visit the potential suppliers before awarding the business. He and a quality engineer took a trip to Taiwan and Indonesia to survey the suppliers.

## *Language Barriers*

Neither the buyer nor the engineer understood Mandarin Chinese or Bahasa Malaysia, the local languages, but the supplier reps assured them that it was not a problem. Indeed, a supplier met them at the airport and took them to a hotel. During the meetings at the suppliers, they saw a pattern of their questions causing a great deal of discussion in the local language, followed by an answer that was much shorter than the discussion. It was obvious that the suppliers were caucusing during the meeting to prepare their answers better. The buyers also left some meetings uncertain what had been agreed to.

They had made a mistake by not bringing their own interpreter to the meetings, and they compounded the mistake by not clearly documenting the meetings before leaving. Chapter 7 gives advice on how they could have handled this better.

### *Cultural Understanding?*

The engineer and the buyer enjoyed their meetings with the suppliers, who were consistently courteous. They felt there were no cultural problems.

However, during a factory tour, they eliminated two perfectly good suppliers because there was no evidence of employee involvement in statistical process control activities. (Their U.S.-based supplier quality system required this.)

They didn't realize that there are many hidden differences between cultures. One of the biggest is that the culture in many countries makes it difficult for educated, professional staff to give too much authority to ordinary workers. The culture also makes it difficult for the workers to accept responsibility and authority. The buying team had run into an unexpected cultural difference.

The team also held to their company's policy of not permitting suppliers to buy dinners or entertainment. This frustrated the suppliers' sales staff, who felt they did not understand the buyers sufficiently to deal directly. The team did not realize the higher importance of a good human relationship between buyer and seller in most countries.

This book encourages as much direct dealing as possible between buyers and sellers. To do this successfully, buyers must recognize cultural differences and deal with them skillfully. This is a pervasive issue. It will be the first issue dealt with in the book, in Chapters 2 through 6.

## DOING BUSINESS

This story could get rather long. Eventually, and much later than expected, the buyer had placed tooling orders (offshore) and orders for the parts themselves. The buyer had learned something about duties in the process and decided that his company would pay all the duties involved directly, rather than including duties in the price. Samples were exchanged, a contract was signed, toasts were raised, and the buyer settled down to enjoy his expected 15 percent savings, not knowing that he could have done much better.

Soon after the first production shipment arrived, the U.S. Customs Service called to remind him that he still owed duty on the tooling, even though it had not entered the United States. Later that week, the seller called to say that he was unable to hold to his promise of fixed U.S.–dollar-pricing because the dollar had weakened too much. The buyer (and his manager) began to believe critics who said that the savings from international purchasing are often an illusion. If they had learned a few key skills before starting, they would have done much better.

## NOT HYPOTHETICAL

All of these stories are true examples of errors made by buyers and design engineers. Fortunately, all of them did not happen to the same team.

### *KEY POINTS*

- ▶ Locating good foreign suppliers takes a different set of skills than locating domestic suppliers.
- ▶ Macroeconomic considerations do not predict countries' success in producing a particular product.
- ▶ Reluctance to deal in foreign currency raises American buyers' costs.
- ▶ U.S. legal practices are often not accepted in other countries.
- ▶ Cross-cultural communication skills are essential for success in international purchasing.

*Chapter Two*

# Cultural Differences
## *Introduction*

Cultural differences have an unusual characteristic. Experienced international buyers see them to be bigger problems than inexperienced buyers do. This is because inexperienced buyers mistake the surface signs of a culture for its important aspects. They may master the surface aspects and not even recognize much more important cultural differences. It's only through after-the-fact analysis of what went wrong in failed deals that experienced international businesspeople realize there are deeper and more fundamental cultural issues.

In setting out to analyze and overcome cultural differences, I realized that I would need to develop a list of the ways cultures differ. If I could do that, I hoped, I could find a few fundamental differences that would classify a country's culture as it affects international purchasing. This turned out to be a successful project.

## WHAT IS CULTURE?

The dictionary definition is that culture is the sum of the understandings that govern human interactions in a society. People learn those understandings at an early age. Often, the understandings have never been examined or even specifically stated.

These understandings come from schools, from ideological sources, and from less formal influences such as parental and peer pressure. Ideological sources include political, economic, and religious or other ethical systems.

These understandings result in two broad areas of differences. One is values, the way people think. The other is behavior, the way people act. Both of these affect the ability to do business between cultures.

In purchasing, a buyer and a seller enter into a relationship, and they transact business. The cultural chapters of this book will focus on the differences that affect the ability of the parties to enter into an effective relationship. They will concentrate on differences that prevent people from communicating with, and influencing the behavior of, people in different cultures.

## *Behavioral Differences*

The most troublesome behavioral differences usually aren't the obvious ones. The obvious differences tend to be ones of superficial manners. Most people involved in international business know, for example, that people in some cultures bow when meeting someone, and others shake hands. While it is good for either party to learn the other party's customs, these issues rarely get in the way, provided that neither party violates a level of manners called *taboos.*

Taboos are very deeply held proscriptions. They generally involve sex, death, or human waste, and can appear in surprising ways. For example, throughout the Middle East, it is totally incorrect to hand anyone anything with the left hand. (The left hand is used for bathroom hygiene.) Members of this culture learned this at a very young age. Violating this taboo will disturb even sophisticated members of the culture.

Fortunately, you can learn the taboos of another country from the most basic reading. No one should travel to another country on business or pleasure (or host a visitor from another country) without doing some preliminary reading on that culture. (*Culturegrams,* mentioned in this chapter's resource list, is a useful basic tool.) This simple reading will save you from committing serious etiquette errors.

More-subtle differences can be larger problems. For example, I used to be constantly puzzled and mildly annoyed in some countries when people answered telephones in conference rooms and carried on unrelated discussions during meetings. In other countries, people were consistently late for meetings and appointments. I automatically applied American behavioral standards and wondered whether these people were being rude or were just disorganized. When I realized that they were neither rude nor disorganized by their own standards, I became much more accepting and comfortable.

## *Value Differences*

A person's behavior is usually the result of applying his or her values in a situation. If values are understood, behavior becomes easier to understand.

I decided to do some basic cultural research to see if I could find the characteristics that differentiate one culture's business-related values from those of another. After skimming a few dozen books, I found two excellent, business-oriented titles that helped to explain much of what I was experiencing. (These books will be introduced in the next two chapters.)

I was then able to recognize and even expect certain patterns of behavior. Simply naming something can take the mystery out of it. If you can name and anticipate the behavior, you can have a strategy for dealing with it, making you much more comfortable.

## SOME WORDS OF CAUTION

Beware of stereotyping. You will not be able to apply generalizations about a country to everyone in it. There is a range of individuals and individual behaviors within each country. Be prepared for them. Japanese businesspeople have a reputation for being formal and reserved. However, you will occasionally run into an extroverted, gregarious, backslapping, spontaneous Japanese businessman. Don't be too surprised.

There is a particular problem in analyzing the culture of the United States. The United States is, and prides itself in being, a multicultural society. The business world is still dominated by northern-European-based cultural styles, but many companies and individuals do not fit this style. You need to have some understanding of your own and your company's cultural style to understand others'.

## IMPORTANCE OF CULTURAL SKILLS

I have been asked why a buyer needs cross-cultural skills. "After all," some say, "we're the buyer; isn't it the seller's job to learn to do it our way?" Possibly it is, in situations where a buyer can get his or her way through sheer economic clout. Very few situations are this simple, however.

## *Cultural Pervasiveness*

Every country's business practices are based on the country's cultural patterns. These patterns grew up over a long period in each country, and while international trade is reducing the differences, they still survive. The following are some of the key areas where cultural differences affect buying.

**Information receiving.**   One of the more surprising cultural differences is that countries vary in how people receive information. In some countries people are used to receiving information in formal briefings. In others, briefings may be only one step in a multistep process of gathering and accepting information. A buyer who tries the wrong technique will get the wrong result.

**Decision making.**   This is one of the most important cultural differences. In some countries, decisions are made in a very top-down fashion. One key individual with a lot of power will make decisions in the company, and subordinates will be expected to carry them out. In other countries, decisions will be made in a more democratic or even bottom-up fashion. In such countries, decisions cannot be made until there has been time for discussions on several levels. A consensus has to evolve, and it is likely to be a consensus not only on the decision but also on how to implement the decision.

**Legal practices.**   Legal practices vary with the amount of trust that is assumed in a business relationship. Countries where people do not automatically trust one another (like the United States, sad to say) will have very complex and thorough contracts and confrontational legal techniques. Countries where trustworthy behavior is assumed will have simple contracts and cooperative techniques.

**Motivation.**   In the United States, we are used to appealing to a person's self-interest in order to get cooperation. In other countries, an appeal to the welfare of a group is likely to be a better motivator. In fact, praising an individual at a supplier in some countries may cause problems for that person. A skillful buyer will know the better way to motivate people and will be able to pick the right technique.

## GETTING TO THE RIGHT PEOPLE

Later in this book, I will be stressing the importance of breaking through the supplier's representative and intermediary structure. You can get the best results by dealing directly with the supplier. One of the reasons why the suppliers set up a sales or rep structure in the United States is to help shield their own managers from culturally difficult situations. If a buyer appears to be difficult to deal with because of a lack of cultural skills, the supplier is less likely to want to deal directly with him or her.

When a supplier is willing to deal directly with you, you will need to take cultural differences into consideration during every aspect of your dealings. Negotiations are different, decision making is different, the factory floor may be organized differently, and the way to persuade your supplier to meet your needs will vary from country to country.

## AMERICANS' REPUTATION

Americans carry an extra burden in cultural matters. We used to be the unquestioned richest country in the world. (We are now in a close race with Switzerland, Germany, and Japan in per capita gross domestic product.) In the past, our travelers abroad developed a reputation for insensitive and ignorant behavior. Some of these travelers were on business and some were on vacation. Some of this reputation remains, and I hope that purchasers traveling abroad will not contribute to that perception.

## *KEY POINTS*

- ▶ Experienced international buyers are more likely to see cultural differences as a problem than inexperienced international buyers are.
- ▶ Differences in manners are usually not the biggest cultural problem, unless taboos are violated.
- ▶ Hidden differences and unexpected differences are larger problems.
- ▶ Knowing where and how cultural differences will appear improves a buyer's comfort and effectiveness.

## *RESOURCES AND REFERENCES*

*Culturegrams.* Garrett Park, MD: Garrett Park Press. A two-volume set of basic information on nearly every country in the world. Each country has three–four pages of information.

Hill, Charles W. L. *International Business: Competing in the Global Marketplace.* Burr Ridge, IL: Richard D. Irwin, 1994. Chapter 3 gives a good summary of cultural concepts.

*Trade and Culture.* Baltimore, MD: Trade and Culture Inc. A quarterly news magazine that has helpful tips about several countries in each issue. It also includes a region-by-region news summary.

*Worldwide Business Practices Report.* Deerfield, IL: Intercultural Enterprises Inc. A monthly newsletter that gives good background on a few countries every month. It focuses on business customs and protocol.

## Chapter Three

# Cultural Values

I had long been wondering whether there are a few fundamental dimensions along which cultural values could be measured. If there are, then different cultures can be placed on a scale and generalizations can be made about them. I was glad to find that a book has been written about cultural value differences that does just that. This book focuses on the values that affect work.

## CULTURE'S CONSEQUENCES

The book is *Culture's Consequences, International Differences in Work-Related Values,* by Geert Hofstede. Hofstede is a Dutch researcher who wrote the book based on research done at a large American multinational company. In the book, he calls the company by the pseudonym HERMES, but it is widely believed to be IBM. He did the research between 1967 and 1973. While some of his terminology may appear sexist today, he still gives very interesting insights.

This book will appeal to analytic minds. It names four key work-related values and makes numerical measurements of the importance these values have in several countries. He clusters countries based on pairs of these values, and shows similarities and differences. If you like graphs and charts, this is a book for you.

### *Power Distance*

The first of the characteristics Hofstede discusses is "power distance." It is a measure of the inequality of power and influence within a society. A high score indicates that there are major differences in power and influence between the more-powerful and less-powerful. This difference may or may not make the subordinate unhappy or rebellious. It's probable that those in the subordinate position view the inequality as a normal and even comforting certainty in their lives.

Exhibit 3–1 plots power-distance scores for several of the United States' major trading partners. This graph shows the United States to be intermediate in terms of power distance. Mexico, which has a strong tradition of small companies run by the owner (the *patron*), is high on the power-distance scale. Israel, where soldiers call their officers by their first names and where it is difficult to teach soldiers how to salute, is very low on the scale.

This data surprises many people because some countries don't seem to match their image. Japanese are, for example, surprised at how low their country is on the power-distance scale. The Japanese language has rank built into its very grammar. There is not one widely used word for "give," for example. There are several words denoting the relative status of giver and recipient. One gives (*ageru*) to a higher-level person, or gives (*kudaseru*) to a lower-level person.

**EXHIBIT 3–1**
*Power Distance*

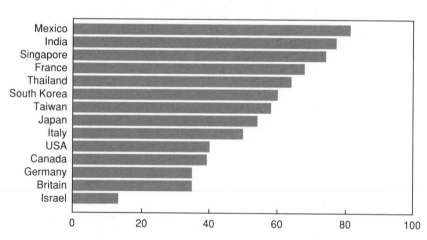

Source: Geert Hofstede, *Culture's Consequences: Differences in Work-Related Values,* abr. ed., p. 77. Copyright © 1980, 1984 by G. H. Hofstede. Reprinted by permission of Sage Publications, Inc.

I believe Japan gets its lower-than-expected ranking for two reasons. The first reason is that there is a relatively level income distribution throughout industrial Japan. The salary spread between industrial worker and company president is much less than it is in the United States.

The second reason is that much of the apparent inequality in Japan is situational, rather than permanent. A flight attendant will treat an airline passenger with great deference. However, if that airline passenger is also a store owner and the attendant enters the store as a customer, the store owner will treat the flight attendant with the same deference. Customers in Japan always outrank sellers in the context of a buying and selling relationship. This is a very different system from the "class structure" of large and seemingly permanent status differences built into many societies.

Germany is another country that has a surprisingly low power-distance score. The tradition of the authoritarian German leader who demanded and obtained complete obedience went into disfavor after 1945.

There is a strong German governmental push to social equality. For example, laws require all stores in the country to close at the same early hours on evenings and weekends. A visitor without a toothbrush flying into Germany on a Saturday afternoon will not be able to buy one until Monday (except at the airport). Part of the reason for these rules is a belief that retail salespeople have the same right to a work-free weekend as the rest of the country.

German drivers used to be famous for making an angry display of lights and horns and forcing slower cars out of the fast lane of an *autobahn*. This practice was recently made illegal because it was deemed to be a show of too much power over others.

There are lessons for buyers here. Displays of power, either personal power or the buyer's power over the seller, will be more or less appropriate depending on the country. In most Asian countries, the customer has higher status and can use that status easily. There will be a strong tendency for sellers to want to meet the buyer's requests. In Germany or Israel, by contrast, such displays will cause strong resistance and resentment.

However, even in a country where buyers outrank sellers or where displays of power are seen as normal, visitors should proceed cautiously until they understand the ground rules. If a country has a tradition of granting every buyer's request, it must also have a built-in set of con-straints on buyer demands or the country would not remain functional. A too-agressive potential buyer may never become a customer.

In addition, the level of power distance will strongly affect the setup of factories. In the United States, we prefer to see workers making their own statistical process control measurements, for example. This does not happen as often in high power-distance countries. Such tasks are reserved for professionals, and not left to factory workers. Buyers should realize that there are many methods of achieving quality, and focus on the results and not the techniques.

---

### Cultural Miscues

I don't think Americans are unusually culturally insensitive. We are a multicultural society and are often aware of cultural differences. There are cultural miscues on the other side of the negotiating table also.

One time, a group of buyers was in Mexico City evaluating a potential supplier. They had a definite time by which they had to leave in order to catch an international flight. The owner of the company, a son of a wealthy and well-connected family, told them not to worry about missing their flight. Finally, when the buyers were sure it was too late, the owner brought a Mexico City police sergeant into the room. He was introduced as their escort to the airport, and in spite of his uniform said to be someone who worked for the company owner. The buyers were given a full police escort to the airport, with sirens and lights used on the expressway in order to clear less-privileged traffic out of the way.

The supplier did not get the business. Most people from the United States are very uncomfortable with such an exercise of power. A more culturally attuned owner would have realized this.

If there is a major cultural failing of Americans, it is a failure to learn the rudiments of other countries' history or geography. A display of ignorance of the world outside the United States reduces the prestige of the American visitor. Ignorance can interfere with business negotiations in countries where personal relationships are crucial.

---

## Uncertainty Avoidance

Uncertainty avoidance is the second of Hofstede's four key cultural differences. This is a measure of how uncomfortable a society is with uncertainty. A high score indicates a strong discomfort, and a low score indicates a low discomfort. Exhibit 3–2 shows the variation between countries.

**EXHIBIT 3–2**
*Uncertainty Avoidance*

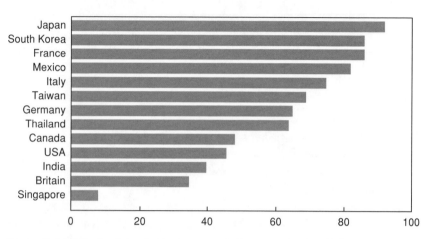

Source: Geert Hofstede, *Culture's Consequences: Differences in Work-Related Values,* abr. ed., p. 122. Copyright © 1980, 1984 by G. H. Hofstede. Reprinted by permission of Sage Publications, Inc.

This graph shows that the United States is moderately low; Britain is still lower; and Singapore, a former British colony, is extremely low. It shows that Asian countries are no more alike than European countries, with a major difference between Singapore and Japan. Significant differences exist within Europe also. France approaches Japan for high uncertainty avoidance.

Signs of high uncertainty avoidance include a high degree of rules and rituals, generally developed as a method of warding off the fears of uncertainty.

**Recent changes.** I disagree with a few of Hofstede's country scores. Some of the scores seem incorrect today because of cultural changes that have taken place in the last 20 years. I believe Singapore has a higher uncertainty avoidance than shown, and I would place it near Canada. The change is probably a result of Singapore's changing from a British-influenced to a Chinese-influenced society. I would also lower Taiwan's score to one near Canada's, and raise South Korea's score above Japan's.

In Europe, I would lower Italy's score to a level below Germany's. Italian industry seems flexible and responsive to me.

**Lessons for buyers.** The lessons for buyers are obvious. Do not expect people in countries with high uncertainty avoidance to respond well to sudden surprises, either in negotiations or in production scheduling. Expect detailed and meticulous planning. Signal changes well in advance, and allow the other party to discuss and react on their own time schedule.

Once planning is complete and production starts, expect limited scheduling flexibility in countries with high uncertainty avoidance.

## *Individualism*

The third of Hofstede's key characteristics is individualism. This is the balance between the needs and wants of individuals and those of larger groups in a society. Exhibit 3–3 shows the United States and many of its major trading partners.

The graph shows that the United States is very strong on the individualism scale, with Britain and Canada quite close behind. The countries of Asia are more group or family oriented. Individualism is the only characteristic for which the United States has an extremely high or low score.

**EXHIBIT 3–3**
*Individualism*

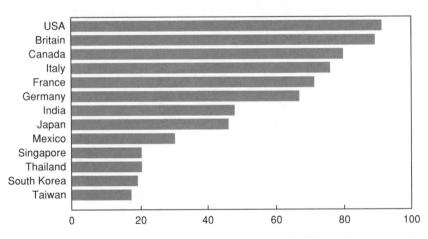

Source: Geert Hofstede, *Culture's Consequences: Differences in Work-Related Values*, abr. ed., p. 158. Copyright © 1980, 1984 by G. H. Hofstede. Reprinted by permission of Sage Publications, Inc.

**Individualism in Japan.**   Hofstede shows Japan to be much higher on the individualism scale than I believe is true. I believe this is partly because the survey was done among employees of an American company and joining such a company takes an unusual degree of individualism.

My experience is that belonging to a group is essential to the makeup of most Japanese to a degree that is hard to realize until you see it. Company employees work together, go on trips together, and even live together in company dormitories.

Compare the U.S. expression, "The squeaky wheel gets the grease," to the Japanese expression, "A nail that sticks up gets hammered down." These expressions describe two distinct cultural attitudes to individual assertiveness and achievement. There is a story that a Japanese person was challenged to find an equivalent to the "hammer" expression in any Western saying. He is said to have found the "squeaky wheel" saying and to have had difficulty accepting that the meaning of the "squeaky wheel" expression was to encourage individualistic behavior.

Again, the lessons for supply managers are plain. Don't expect a single individual to make fast decisions as you would in the United States. In addition, U.S. supply managers should restrain tendencies to want to appear as the "lone cowboy" decision maker, willing and capable of making any decision on the spot. Such behavior will disturb and concern people in many countries. Even worse, negotiators might take advantage of the situation by making it difficult for a single American negotiator to contact his peers or supervisor for advice, hoping he will make an uninformed decision.

In countries with low individualism scores, concern for groups should go on well beyond the negotiation and initial business steps. Think in terms of praising groups and teams at the supplier, rather than individuals. Mention more often that groups at your company are pleased or concerned. Try not to appear too individualistic yourself.

### *Masculinity*

This is Hofstede's term for the fourth dimension of cultural differences. It is an unfortunate term, I believe, because it contains connotations that interfere with discussion of the issues he raises. Others have named it "performance orientation," as contrasted with "empathy."

He believes that assertiveness is a masculine trait, and nurturing behavior is a feminine trait. He has data from the 1960s and 1970s to support that conclusion. His measure of masculinity indicates two things: how well women are integrated into work life and how much organizations adapt themselves to supposedly feminine traits.

I have not seen much relevance of this trait to international purchasing. However, I'm including Exhibit 3–4, which shows the scores, because I find it correlates with how well a culture accepts its own women into business life. A high score shows that they are not well accepted, and a low score shows they are.

This graph also illustrates another major difference among Asian cultures. Japan has a much higher masculinity score than the southern Asian countries of Taiwan and Singapore. I have also observed that women in northern Asia (Japan and South Korea) have much more difficulty being accepted into business and public life than women in southern Asia (Taiwan, Hong Kong, and Singapore).

If a high score indicates low acceptance of the country's own women in business, then South Korea deserves a higher score than Japan.

**EXHIBIT 3–4**
*Masculinity (Assertiveness)*

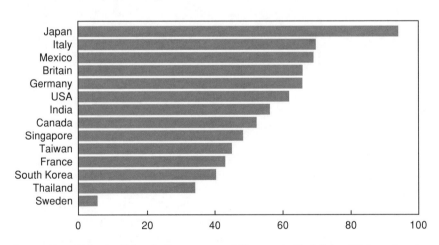

Source: Geert Hofstede, *Culture's Consequences: Differences in Work-Related Values,* abr. ed., p. 189. Copyright © 1980, 1984 by G. H. Hofstede. Reprinted by permission of Sage Publications, Inc.

**Women travelers.**   Women business travelers going to Asia for the first time often worry about their reception and treatment. I believe this concern is largely misplaced. Foreign women business travelers are generally seen in Asia more as foreigners than women. The behavior adjustment required to deal with a foreigner is so great that dealing with a foreign woman requires almost no more effort or concern. As more and more foreign visitors arrive, however, this perception could change. Foreign visitors will no longer be seen as rare or exotic.

In contrast, some European countries are, of course, familiar with foreigners but are not accustomed to having women in senior business positions. Treatment will be polite, but there may be an excessive concern for the supposedly delicate status of a woman that can appear condescending.

In either case, if the woman is the senior person on a traveling team, men accompanying her should go out of their way to make her rank apparent. This includes giving her the seat of honor, deferring to her in meetings, and treating her more formally than others.

Women business travelers will not be accepted in traditional Arab countries. It is taboo for men and women who are not related or married to mix, even in business. American companies should carefully consider whether buying from those countries is essential to their competitive strategies. While we should not impose our cultural standards on other countries, neither should women employees be denied a full role in our companies.

## OTHER VALUE DIFFERENCES

In addition to Hofstede's four differences, I have identified two more that I think are crucial.

### *Need for Harmony*

In most Asian cultures, there is a much stronger need for harmony than in Western cultures. This strongly affects decision-making practices. There will be a tendency to be sure that the needs of all members of a group have had proper respect and consideration. People will seek compromise over confrontation. A confrontational negotiator will not succeed in such an environment.

On the negative side, this can lead to a supplier's stating things that are not true in order to preserve harmony. This happens in the United States also, but usually in an effort to spare the feelings of a person about a personal matter. For example, very few Americans would answer no to the question "Do you like my new clothes?"

There will be more on the topic of saying no in the next chapter.

### Guilt, Shame, and "Face"

Some religions have a concept of guilt that other religions do not have. If a person belonging to a religion that stresses guilt acts in a way that he or she believes is bad, this person is likely to feel uncomfortable even if other people are unlikely to become aware of the behavior. For these people, an omniscient God is thought to know all, and this is a powerful restraint on behavior.

Shame, on the other hand, is the reaction to other people's perception of one's behavior. It is more likely to be the dominant value in low-individualism countries, where children are raised to give prime consideration to the needs of family and neighbors. Given the relatively low crime rates in those countries, shame is also a powerful restraint.

People in shame-driven countries are going to be much more sensitive about being embarrassed in front of their peers. This leads to the concept of "face" and explains why it is more important in many Asian countries than it is in the West. To some extent the approval of peers is as important as the approval of a God is to religious persons. Be sure to proceed with great caution when there is any possibility of embarrassing an individual in a country that is low on individualism.

### KEY POINTS

- ▶ Geert Hofstede identified four key work-related cultural values: power distance, uncertainty avoidance, individualism, and masculinity.
- ▶ Other key value differences that affect buyers are: drive toward harmony, buyer-seller rank, and orientation toward guilt or shame.
- ▶ Buyers should adjust their expectations and behavior according to the values of the seller's country.

## RESOURCES AND REFERENCES

Hofstede, Geert. *Culture's Consequences: International Differences in Work-Related Values,* abridged edition. Newbury Park, CA: Sage Publications, Inc., 1984.

*Chapter Four*

# Communication Differences

The most subtle behavioral differences are those involving communication style. Unexpected differences in communication and decision-making style can reduce your effectiveness in influencing others.

Edward and Mildred Hall's *Understanding Cultural Differences: Germans, French and Americans* provides some very useful insights. The book concentrates on differences in communication style. While it focuses on Germany, France, and the United States, it provides insight into more general communication style differences. This book is less analytical. It relies more on intuition, and is much easier to read than the Hofstede book. The two major communication style differences the Halls mention are message speed and level of context.

## MESSAGE SPEED

People in different countries are used to having different amounts of time to receive information, process it, and respond. There is a normal speed of messages. In the United States, a television commercial is a fast message. A *Consumer Reports* magazine product review is a slow message. A newspaper headline is a fast message. The accompanying newspaper article is a slower message, and a book on which the article is based is a still slower message.

The U.S. communication style is extremely fast. Conversation and writing tend toward "headline" style, with the conclusions first. Long reports will have an executive summary at the start, and many readers will not go beyond that unless they disagree with the conclusion. Americans usually mention the most important points at the start of a conversation or verbal report. People in other cultures, especially those of northern Europe, tend to take more time to build cases and reach logical conclusions.

An American runs a real risk of not getting his or her message across if he or she is using U.S. communication styles in other countries. The

American is likely to make a statement that is a conclusion. If no one disagrees or asks questions, the American will move on and not give any explanation. The American will take the other party's silence as a tacit agreement. The other party may think the American is shallow and illogical.

I haven't seen a general classification of countries by communication speed. The best way to determine the speed is by listening to the communication style of the other party. What comes first, the conclusion or the facts supporting that conclusion?

I believe that communication style varies within countries also. U.S. academic writing tends to start with facts and build to a logical conclusion. Business writing is the opposite; the point of the message is first. In Japan, television commercials are usually only 15 seconds long, but business discussions are usually much longer than they are in the United States.

---

*Communication with Germans*

I found difficulties communicating with my German fellow workers, and I had to change my style to overcome them. I found that many of the Germans would, while talking, speak in flawless sentences that flowed into well-crafted paragraphs, which would lead to a good logical conclusion.

I might have disagreed with one or two of the facts mentioned in an early part of the discussion. However, I had been trained as part of my culture not to interrupt people. That's an easy rule in the United States; people rarely talk for a long time. However, as a German continued to talk, I would find myself so impressed that he or she could speak in English more logically than I generally do, that I would tend to forget what my objection was by the time he or she finished speaking.

I finally realized that I either had to take notes or interrupt at an early point in the discussion. Sorry to say, I didn't always have a notepad with me.

---

## HIGH- OR LOW-CONTEXT

The second major variation between communication styles is the degree of context. "Context" refers to the amount of information that is NOT communicated explicitly in a message because the sender assumes it is already known.

With a high-context communication, the sender tends to assume that the recipient already knows a lot of background information. A low-context communication contains a lot more information because the sender assumes that the listener does not know very much of the background.

An encyclopedia is an example of a low-context communication. Stock tables and sports articles are examples of high-context communication. They are full of jargon, acronyms, and abbreviations. Only a reader who knows the background can understand them.

Communication styles reflect the lifestyle of a country. Japan, for example, is a country where people are constantly moving in networks of friends, relatives, fellow workers, suppliers, and customers. They are in frequent contact with all those people and know what is happening. Workers in a high-context country are said to be "swimming in a sea of information." In turn, Japanese communication style is very high-context. Frequently the subject of a sentence cannot be determined without knowing the context behind the sentence.

France is similar to Japan in that direct, frank speech is not considered sophisticated. Clues, hints, and allegories are used to get the message across.

Germany, on the other hand, is very low-context, and a great deal of explanation is necessary. This major difference in communication styles that exists across the French–German border can lead to a great deal of confusion and misunderstanding.

Americans tend to be fairly high in context, but not as high as most northern Europeans. We generally do not "network" well. We have been mobile in living places and careers for a number of generations, which can cause us to believe that relationships are transient.

When you are having a discussion with a person from a different culture, you need to be aware of both of your normal context levels. Communicating at too high a level of context (assuming the listener or reader knows a lot more than he or she does) first confuses and ultimately frustrates the receiver. He or she will not understand you at first, and will usually first attempt to draw out information with questions. If unsuccessful, he or she may just give up. Communicating with a person from a higher-context culture who speaks another language is particularly difficult, because communication problems caused by context may be mistaken for language problems.

In contrast, communicating at too low a context (explaining things a person already knows) will initially irritate and finally bore the other party. I have heard such communication referred to as "explaining the obvious with great sincerity."

There are implications for communication inherent in the level of context. Information flows freely in high-context societies and organizations. It doesn't flow well in low-context societies, which tend to be more compartmentalized. In communicating to low-context organizations, speakers will have to start over every time new groups join the action. The best strategy for dealing with this is to take written notes covering everything that has been presented and discussed. You can request new parties to read the notes ahead of discussions.

Usually, Americans will be communicating into higher-context societies. People in those societies are used to a constant flow of information from a variety of sources. They usually know these sources well. They are probably not used to learning everything at a formal presentation or by reading one paper. In these cases, be sure to allow for extra time for informal communication, generally outside what we in the United States would call a working environment. They may want to retire to a bar and consider what you said over a few drinks. You should also expect that they will want to get to know you as a whole person, not just your business role.

They will certainly want to be sure they understand their co-workers' opinions of material you presented. Asking for agreement or disagreement directly and quickly after a presentation will not work. Look for signs of concern or lack of understanding and try to draw out the differences. Obviously, discussion takes more time in these societies.

### Country Classification

The Halls mention Japan, Mediterranean Europe, and Arab countries as high-context societies and the United States and Northern European countries as low-context. They do not attempt to classify other major trading partners, such as other Asian countries.

You will need to look for clues, and one major clue is office layout. Are all co-workers grouped together, with limited privacy and the likelihood of everyone hearing what is happening? Is the boss's office or desk right with the work group? These are all signs of high-context societies.

On the other hand, are there long halls with private offices on both sides? Are the doors generally opened or closed? Barriers, walls, and high degrees of privacy are signs of a low-context society. If you are not sure what to expect, bring both high- and low-context communication materials so that you are prepared for either approach.

The Halls refer to the United States as a low-context country, but there are significant differences within the United States and among its institutions. Many U.S. companies have done away with private offices. They use low-walled cubicles that do not give a great deal of privacy. People are more aware of what is going on around them.

Information seems to flow a lot more readily in the companies that have cubicles. I believe that organizations that have a lot of private offices not only suffer from poor communication, but they are also more prone to rivalries and unrestrained internal competition and conflict. You should consider the level of context in your own company, rather than assume that you are from a low-context society.

## GETTING TO "NO"

One of the most frustrating experiences for negotiators is to believe there is agreement but to find out later that there was none. This can happen for easily avoidable reasons, such as mistaking a "yes" for agreement when it really meant "I hear you." This happens to novices occasionally in Japan, but this linguistic issue is becoming better and better understood.

Bigger and more difficult problems occur because people in some countries find it difficult to say no. This is often hard for Americans, who value frankness and openness, to understand. Often, reluctance to say no is a result of a drive for harmony.

Communication within a culture that strives for harmony requires restraint and sensitivity by both parties to a negotiation. In order for such a culture to function effectively, it is also necessary for a person not to put another in a position where he or she will have to say no.

Linguistic patterns can make this easier. In Japanese and some other Asian languages, it's easy to construct a sentence with no sign whether it will be positive or negative until the last syllable of the last word. Consider a language where the following is correct word order: "With respect to that person's proposal, I agree not." The speaker could watch body language and other clues as the sentence progressed, and alter the total meaning of the sentence at the end to suit the expectations of the listener. No disagreement would be obvious, and surface harmony would be maintained.

If a person from another culture does not understand this trait, and forces a negotiator into a corner where he or she must say no bluntly and openly, the results are unpredictable. The negotiator may, in the interests of harmony, decide that it is not wise to state the truth openly.

In these countries, which include Asian as well as some Latin American countries, it's better to ask open-ended questions, such as "What is the most difficult part of our proposal?" rather than "Do you like our proposal?" This will enable the other party to say, "Your proposal, overall, is excellent, but perhaps we could work together more on this particular section."

## KEY POINTS

▶ The two largest differences in communication styles are message speed and level of content.

▶ Americans generally give fast messages, with the conclusions expressed first. This style of communication is inappropriate in many countries, particularly Europe.

▶ High-context communications assume the receiver understands a lot already. Societies oriented toward high-context communications will have a lot of networking and informal communications. This process requires time.

▶ To communicate effectively with high-context suppliers, it is important to allow time for them to understand and accept you and your company.

## *RESOURCES AND REFERENCES*

Hall, Edward T. and Mildred R. *Understanding Cultural Differences: Germans, French and Americans,* Yarmouth, ME: Intercultural Press Inc.

*Chapter Five*

# Other Behavioral Differences

There are other key behavioral differences that are not related to communication.

## SPACE, TERRITORY, AND PRIVACY

The Halls' book points out this behavioral difference. It includes simple issues such as how close people stand when talking, and more complex issues such as how people perceive possessions and how much they compartmentalize their lives.

### Space

The distance that people stand from each other is a subtle cultural difference that sometimes causes discomfort until people recognize what is happening. People from Mediterranean (and Mediterranean-colonized Latin American) and Arab cultures tend to stand closer and touch more than people from northern European cultures.

Northern Europeans and most Americans tend to give each other a wider space. Japanese also prefer a wide physical distance from each other and do not particularly enjoy being touched by a new acquaintance.

As usual, Americans are a complex group. First, we have the northern European tendency to avoid touching and to avoid close physical proximity. On the other hand, many people see us as too familiar and too quick to make friendly verbal and social gestures and strive toward an apparent intimacy. Calling people by their first names is an example. In most countries, this is reserved for relatives and long-standing acquaintances. We are sending mixed signals. We are signaling that we want to be friends by our speech but signaling hesitancy and reservations by our physical behavior. This can be perplexing to others.

## PRIVACY AND TERRITORY

Countries vary widely in their views of what is private and what is public. In some countries, it's rude to ask about a man's wife and family. In other countries it will be considered rude not to ask.

Look for clues, such as whether the other party wants to get down to business immediately or whether he or she seems more interested in learning about your family and your personal interests first. In some countries business will be done only after a relationship is established. In others, the relationship might or might not develop as business proceeds, depending on whether the other party regards you as worthy of friendship.

### *Compartmentalization*

In this context, "compartmentalization" refers to a tendency to divide lives into business and personal areas. The difference is deeper in some countries than in others. Asking or suggesting that people in some countries work outside normal hours might cause strong resistance. Even a social time with a boss, client, or customer may be seen as work.

This tendency seems strongest in northern Europe. In most of Germany, companies must obtain approval from a local government to have employees in the building on a Sunday. Japanese and most other Asians are less likely to have firm borders between work and private lives. Work is just another part of life.

In working with people in other countries, be sure to take this into account. Asking someone to work on a weekend might cause offense in some countries. On the other hand, if someone from a highly compartmentalized country invites you to his or her home, it is an honor indeed and a sign that you have gained respect.

## TIME

This is a very obvious cultural difference and one you will instantly notice. According to the Halls, cultures are either monochronic or polychronic. Monochronic people and cultures like to do one thing at a time without interruption and with heavy attention to schedules. Polychronic people, on the other hand, like to do many things simultaneously and like to pay more attention to completing the current human transaction than they pay to schedules.

## Monochronic Cultures

The United States and most northern European cultures are monochronic. Time is something that is "spent" or "wasted." We have a saying, "Time is money." Most professionals have a desk calendar that parcels out their day in 15-minute increments. Many have two, one of which they carry with them so that they can always answer questions about their schedule and availability. Being on time is so ingrained in our society that it seems natural and normal, and anyone who is late is considered "disorganized" or even rude. Monochronic people concentrate well and adhere to established plans.

Monochronism is not "natural" but a product of long years of industrial scheduling practices. Farming communities and cultures must be polychronic to survive. Farm animals do not conform to schedules very well. They get sick, they have offspring, they do unpredictable things. Weather is also unpredictable, and a sudden rainstorm will make the most careful scheduling irrelevant. However, the industrial revolution caused a change in that factory workers had to be at work at a set time, and a great deal of training and conditioning resulted. In San Francisco, even today, shipyards blow steam whistles that can be heard all over town to announce the start and end of work.

## Polychronic Cultures

Polychronic cultures and people are characterized by involvement with several things simultaneously and by a heavy emphasis on human interactions. People are not late because they are "disorganized." They prefer to finish a current human interaction even if it means being late for the next meeting. To do otherwise would have seemed rude.

According to the Halls, polychronic cultures also tend to be high-context, which means that their style of decision making is consensus-driven. They need time to discuss things and the decision-making speed may appear slow. This is probably true in some areas, but there are authoritarian (per Hofstede) polychronic cultures such as Mexico, where consensus is less of an issue.

*Cultures and Time*

Being familiar with the Halls' work has helped me to understand our own and others' cultural values better. Once, in Germany, I was scheduled for an early morning meeting with my own employees, followed by a company council meeting with attendees from all over Europe. When I got to the employee meeting, I found that this particular group was very upset with some of the working conditions, to the point where they were ready to look for another job inside the company. I stayed and worked with them, and as a result I was late for the council meeting. The German running the meeting was visibly upset. However, I had felt it more important to complete the human transaction with my employees than to adhere rigorously to schedules, a classic polychronic action.

In another case, I was with my American boss and the manager of our Mexican purchasing office visiting a potential supplier in Mexico. This was our last stop on the way to the airport, and we were taking a plant tour. The supplier's people asked us to see "one more thing," and my boss, the Mexican manager, and I all checked our watches simultaneously. We had all been exposed to the Halls' work, recognized this as classic monochronic behavior, and immediately started laughing at ourselves. So did the supplier. Our Mexican manager said he had been dealing with *norteamericanos* so long he was starting to act like one.

I believe that cultures that tend to stay polychronic are those where the infrastructure is not highly evolved. Mail may not be delivered on time or at all. Factory deliveries may be late due to poor roads, a rail strike, or other events and conditions that make rigid scheduling a futile exercise. This can lead to a sense of pessimism. On the brighter side, it can lead to a marvelous amount of flexibility and creativity, such as in Italy.

Some cultures are a puzzling mix. Japanese trains, production schedules, and appointment keeping are all completely monochronic. However, the planning process seems to be run on a basis of "take the time that is necessary to get it right." Obtaining a rigid development schedule can be difficult.

## Which Is Better?

This is a good example of the risk of trying to decide whether one culture is "better" than another. There are advantages to both kinds of cultures.

Buyers need deliveries timed to meet customer needs, regardless of whether cultures are polychronic or monochronic. Monochronic cultures are likely to have longer, less flexible lead times. Orders will more likely ship on the agreed-upon date. However, by the time the date arrives, your needs may have changed. This can be even more frustrating than the short lead times and occasional lateness of a polychronic culture. Long lead times lead to high variation in order rates to avoid either excess inventory or out-of-stock situations.

## CULTURAL ASPECTS OF TIME

Besides monochronism and polychronism, the Halls discuss lead time, synchrony, and appointments. The appropriate lead time for a meeting will vary from country to country. In any society, the most important events are granted short lead time. Schedules will be dropped and torn up for an important visitor. What differs is what is chosen as important. In some societies, the most important visitor or situation is selected based on personal relationships. In others it is selected based on the position, power, or authority of the visitor. Misunderstandings can result when visitors from one type of society attempt to schedule a visit in other societies.

Synchrony means how attuned speed expectations are between people from different cultures. How long do items take to get done, or even started? My personal "speed" measure is how long it takes to have a shirt laundered. No matter where you are, it takes about one to two hours to wash, dry, and press a shirt at a laundry. However, in the United States, it takes about three to five days to get done. In Japan, it takes one day. I predicted that service in northern Europe would be even slower than in the United States, although I was unable to confirm this. The answer, I was told, was "Our wives do our shirts."

The key issue is how big people's physical and mental in-trays are. There is a culturally appropriate time to let tasks sit before starting. In the United States, items that require hours of activity to complete generally take days to actually get done. Items that require days take weeks, and sometimes more. This is changing gradually due to increased consumer pressure, but it is still the general rule. These expectations of appropriate elapsed time vary from country to country.

## APPOINTMENTS

Punctuality can be another source of friction. In the United States, it is rude to be late for an appointment. It's even ruder in Germany. In other countries, it is not necessarily rude or an insult. Possibly there was a traffic problem, or possibly there was a human transaction that was essential to complete beforehand. Skillful global supply managers will not project their own cultural values onto others when someone is late for an appointment.

## EASE OF COMPLETING ACTION CHAINS

The last of the Halls' list of important cultural characteristics is the ease with which action chains can be completed. An action chain is a series of events that must all be completed to get to a goal. According to the Halls, people from monochronic cultures are significantly affected by interruptions. Once distracted and taken off the schedule, they find it hard to get back on track. People from polychronic cultures, on the other hand, are likely to interrupt themselves, try something new, and then get themselves back on track. They are more flexible and will react better to a change in plans.

The Halls also point out that conflict resolution or escalation is an action chain that varies from country to country. Every culture has its "escape hatches." These prevent disputes from getting out of hand. Someone from the outside may not know the cues and signs showing escape routes and become angry or upset at feeling trapped. On the other hand, he or she may not see that the other party is trying to minimize a conflict and may press on too aggressively.

The classic American escape hatch is to tell a joke when conflict starts escalating. In Japan, particularly between boss and subordinate, the escape hatch is to go and get drunk. They suspend the rules of deference, and the subordinate gets to express himself openly. They may never admit to remembering the discussion, but there will likely be an improved relationship.

## GIFT GIVING

Many Americans find giving and receiving business gifts awkward. The practice is much more ingrained in many cultures than it is in U.S.

business. Purchasers tend to see receiving a gift as verging on accepting a bribe, and rarely are they in a position of being expected to give a gift. In other cultures, the flow of gifts between buyer and seller may be two-way, as a sign of their mutually dependent relationship.

In the early stages of business, don't be too worried about giving or receiving gifts. You need not give anything, and if you receive anything, it will be no more than an appropriate souvenir of your visit. If you do want to reciprocate, you can give a similar present. A pen set, or a book about your city or state is appropriate.

You should be aware of some points of gift-giving etiquette. In Asian countries, gifts are not opened in the presence of the gift giver. Thank the giver kindly, give a present back if you wish, and take your leave. Send a thank-you note later. Asian cultural traditions regarding wrapping are intricate. It's best to have your hotel wrap the present for you in an appropriate local style. If you don't have presents for everyone, give one to the senior person only.

In later stages of business, at points when it is appropriate to demonstrate trust or gratitude, more serious presents can be appropriate. You'll need cultural advice on this point, a topic that a later chapter will cover.

## *KEY POINTS*

▶ Different societies have different attitudes toward the relationship between work and personal life. Expectations for individual flexibility toward long hours, weekend work, or calls at home must be adjusted accordingly.

▶ Societies and individuals can be polychronic or monochronic. Polychronic people and societies can do many unrelated tasks simultaneously, are more flexible, but often are late. Monochronic people and societies do one thing at a time, adhere well to schedules, but are inflexible and intolerant of changes.

▶ Societies will have different expectations of appropriate lead time.

## *RESOURCES AND REFERENCES*

Engholm, Christopher. *When Business East Meets Business West.* New York: John Wiley and Sons, 1991. An excellent summary of business etiquette and practices in 10 East Asian countries.

## Chapter Six

# Overcoming Cultural Differences

Errors and miscommunication caused by cultural misunderstandings can add a lot to your procurement costs. They can prevent you from coming to a worthwhile agreement with a supplier who would give you a competitive advantage. They can cause you to use an expensive intermediary to avoid problems. They can cause you to negotiate a poorer deal than you would have otherwise.

These misunderstandings are by no means one-sided; either the buyer or seller can create them. At least one party must have the skill to recognize what is happening and to correct the situation. Here are some important items to keep in mind.

## THE SIX KEY DIFFERENCES

Of the differences discussed in the previous chapters, I think six are key differences that affect buyers and sellers, and I will add two more specifically for East Asian countries. The six key differences are:

- Power distance.
- Uncertainty avoidance.
- Individualism.
- Level of context.
- Polychronic/monochronic time.
- Buyer/seller rank.

Appendix C at the end of this book will show how the United States and major importing countries rank in these areas, and will give some specific hints for dealing with the differences.

There are two other key differences to keep in mind in dealing with Asian countries:

- The importance of harmony in Asia.
- The importance of "face" in Asia.

## KNOW YOURSELF AND YOUR CULTURE

To deal successfully with other cultures, you have to understand the unique characteristics of your own. The United States has a wide variety of cultures, but the usual business culture is:

- Monochronic.
- Medium- to low-context.
- Low in uncertainty avoidance.
- Medium-low in power distance.
- Very high in individualism.

Recognize that other cultures that work perfectly well operate differently. The most likely differences are:

- They will be much less likely to allow one person to make a decision on the spot (lower individualism).
- They will want to be more certain of the future (higher uncertainty avoidance).
- They will be used to obtaining information from informal, familiar networks rather than from formal presentations (higher context).
- They will be less concerned with personal schedules (more polychronic).
- They will be more likely to make public displays of power (higher power distance).

However, there are major exceptions to these guidelines. Germans and most northern Europeans will be quite concerned with schedules and will be used to receiving information through formal presentations. The Germans will be less likely to make public displays of power.

You should also remember that it is too easy to stereotype a country. Individuals in each country will vary from the stereotype. Take time to get to know the people you will be dealing with.

Also consider some of the stereotypes that others have about Americans.

## STEREOTYPING AMERICANS

One of the most common stereotypes is that we are extremely open and apparently quick to want to make friends, but we never seem to achieve the deeper intimacy that people in other cultures achieve in their friendships. We will usually get on a first-name basis with new business acquaintances almost immediately, but even after long relationships we find it difficult to discuss feelings or other matters we feel are personal.

The reasons for this are deep-seated and not easily changeable, but if you have those tendencies, recognize them. Some people in other countries will welcome your friendliness, but others will reject it if it comes too early. If your friendliness is reciprocated, don't take offense if you are asked surprisingly personal questions.

The second major stereotype is that we lack knowledge of other countries. This is probably as true as any other stereotype. We are a large, remote, self-contained country. Imports and exports totaled 16 percent of our gross domestic product (GDP) in 1990, compared to 52 percent in Germany and 81 percent in Taiwan. (Japan, another relatively self-contained country, imported and exported 23 percent of its GDP.) The average American has relatively little business contact with people from other countries.

Our education system reinforces this stereotype. Many of our public high schools are weak compared to other countries' schools. Our higher (university) education is excellent but often narrow. An engineer, for example, can graduate from a university knowing very little about history, even less about art, and absolutely no foreign languages. This would not happen in Europe.

It's easy to counteract those stereotypes. A few hours reading about a country can teach some basics. You can and should learn a few items that a country is particularly proud of in its culture or history. This may be art, architecture, or some political or economic advancement. If anything is taboo (such as left-handed passing of an object) you should know it, and you should learn a few words in the local language, such as "yes," "no," "please," and "thank you." Any well-written tourist guide will cover these issues.

## HOSTING AND BEING HOSTED

You will probably receive offers of social entertainment or sightseeing for weekends and evenings when you are visiting another country. If your company has no rules against accepting such offers from potential suppliers, you should accept. This will help to build the relationship with the supplier that is so important in other cultures.

If your organization has rules against accepting such social entertainment, you will be handicapped because the supplier will feel frustrated in his or her attempts to get to know you. Be sure to let them know that there is a company policy that you must follow.

There is one caution. Entertainment and late nights can be a tactic to wear down negotiators. If you find that there is nightly entertainment and that supplier personnel are trading off nights so that you are always negotiating with a more rested person, be cautious. If you are with a large group from your company, trade off nights also, giving "work requirements" as the reason. This allows some social contact but keeps a fresh team on hand.

You should also try to host foreign visitors when they are in your country. Local sightseeing, an invitation to a sports event, or even a visit to your home are appropriate. New or infrequent visitors to the United States will especially appreciate this gesture.

## CULTURAL ADVICE

First, don't be embarrassed to admit you need advice. Cultural differences can be puzzling. Some simple advice can prevent you from making embarrassing social errors or help you interpret some behavior that you find unusual.

Second, do not rely on the U.S. representatives of your suppliers to provide that advice. Later, as we discuss purchasing channels, the reasons will become apparent. There are two better choices for cultural guidance.

The first choice is to talk to employees of your own representative or subsidiary in the other country. They are familiar with your company's culture and the most common differences, and they can be on your side in negotiations and discussions.

The second choice is to take a one-day course in a specific culture. There are a large number of organizations that do this, especially focusing on Asian cultures and particularly Japan.

## FINAL ADVICE

You will undoubtedly find that you commit some social error along the way. People from other countries will likely do the same to you. Accept either in good grace, without undue embarrassment, because you are neither the first person nor the last to learn something new the hard way. Treat it as a learning experience and move on.

## *KEY POINTS*

▶ There are six key cultural differences that affect buyers in every country, and two more that are of particular importance in Asia. The six world-wide differences are power distance, uncertainty avoidance, individualism, level of context, monochronism versus polychronism, and buyer–seller rank. The two differences largely pertaining to Asia are the importance of harmony and the importance of "face."

▶ Usually, Americans will face societies with lower individualism, higher uncertainty avoidance, higher context, more polychronism, and higher power distance.

▶ Americans should slow down tendencies to move quickly to a first-name basis and should learn something about a country's history and taboos before visiting.

▶ Everyone needs cultural advice. Try to get it from your own company's employees rather than supplier representatives in the United States.

## RESOURCES AND REFERENCES

Axtell, Roger. *Do's and Taboos around the World.* New York: John Wiley and Sons, 1990. This book discusses social etiquette.

Kato, Hiroki, and Joan Kato. *Understanding and Working with the Japanese Business World.* Englewood Cliffs, NJ: Prentice Hall, 1992. An excellent book by a bicultural couple who understand both sides of cultural differences.

# Chapter Seven

# Language

Another obvious complication in dealing with a foreign supplier is that they usually speak another language. Even if they do speak English, it's likely to be the English taught in England, which can differ in unexpected ways from the American language. Fortunately, English has become the international language of business, and most of the contacts you will make overseas will be with companies where the key people will be able to communicate in English.

This is really fortunate for those of us who have difficulty learning foreign languages. However, we need to keep in mind that anyone who is taking the trouble to speak to us in our language, no matter how strongly accented or difficult to understand, has taken a giant step toward meeting our needs. These people deserve our thanks and appreciation—and more.

It's not possible to eliminate the problems caused by language differences. It's inevitable that discussions and negotiations will take longer. However, you can minimize problems by using some or all of the methods described in this chapter.

## COMMUNICATION IS YOUR RESPONSIBILITY

I used to make a statement at the start of a meeting that anyone who was having trouble understanding my English should tell me immediately. I found that this doesn't work because it requires someone to embarrass himself or herself in front of co-workers. It's better to take positive steps yourself.

If a supplier is holding meetings in English, they deserve more than a "thank you." You should take on the responsibility to ensure that communication between the companies is clear. This is not the time to pull "buyer's rank" and put the burden on the supplier. While this may seem to be a lot of extra effort, if you handle it properly you can gain negotiating advantages.

## ADJUST YOUR SPEAKING STYLE

If you are talking or making a presentation to a group for whom English is a second language, you need to make a few adjustments to be completely sure that they understand you.

### *Slow Down*

This is the most important change. Speak much more slowly than normal, and pause between sentences. If you are traveling in a group, it's a good idea to designate one member to be the "speed cop" who will signal to anyone talking too fast. It's really easy to get caught up in a discussion and increase talking speed to your normal level. An alert team member can slow you down.

### *Use Extra Presentation Graphics*

More people will understand written English than spoken English. If you are doing a planned presentation, use more slides than normal, and bring "backup" slides for some of the topics mentioned but not clarified in your talk. Bring some blanks also, so that you can create written explanations quickly as you need them.

### *Write Down Big Numbers*

Asian cultures that base their writing system on Chinese characters have a different numbering system than Western countries. Large numbers in Western countries are divided by mental or real commas every three digits. A million is a thousand thousands: 1,000,000. A billion is a thousand million. Chinese-based writing and speaking systems divide numbers into groups of ten thousand (*man* in Japanese). Their mental commas come every four digits. A million is a hundred *man,* or a hundred ten thousands (100,0000). Fifty thousand is 5 ten thousands (5,0000). The mental gyrations required to convert between the systems are difficult for even the most experienced translators, and errors are frequent. Large numbers are often off by a factor of 10 or 100 after translation. It's safest to write down anything over 10,000.

## *Watch Your Language*

You should not only avoid profanity, but watch out for jargon and acronyms. You are probably aware when you are using your own company jargon or internal acronyms, but you should also be alert to common American figures of speech that will not be clear elsewhere. The term "G-2" puzzled fluent English speakers in my Japanese office. "Ballpark estimates" will be meaningless in countries where people don't play baseball. "Punting" is a U.S. football term that has to do with boating in countries where they speak British English. Sports and military metaphors appear to have infiltrated our language more than we notice.

Jokes are really risky. I don't even try them anymore. I've faced too many blank stares. Humor rarely translates across cultural lines. This includes attempts to defuse a tense situation by making a remark intended as a joke. A sincere statement is better.

## *Watch Your Grammar*

A few aspects of English grammar are especially confusing to many foreign speakers. One problem occurs with negative questions. Consider the following exchange between an American and a Japanese:

American: "Aren't you going to fix that?"
Japanese: "Yes."

The question "Aren't you going to fix that?" is a negative question. Literally it means "Are you not going to fix that?" In Japanese, if the answer to a negative question is "yes," that indicates agreement with what the speaker said—in this case, that they are not going to fix it. A "no" answer indicates disagreement with the speaker—they are NOT "not going to fix it." They will fix it.

The potential for confusion and uncertainty is unlimited. If you get the answer you expected, you will probably wonder for days if the supplier was answering in the Japanese or English mode. My Japanese secretary called time out whenever I asked her a negative question. Fortunately, you can easily rephrase a negative question into a positive one. "Are you going to fix it?" is a better question.

If you are using a sentence with an "if" clause in it, put the if clause first. "If you continue to ship late, we will reduce your contract volume" will be clearer than "We will reduce your contract volume if you continue to ship late." The first way matches Japanese and Korean sentence structure and is more likely to be understood.

### Watch Your Hands

Hand sign language differs from country to country. What is a perfectly friendly gesture in one country may be rude or obscene in another. In England, two fingers with the palm toward the viewer simply means two. But turn the hand around so that the back of the hand is showing, and you have made an obscene gesture. (Maybe that is why those two beers never arrived.) Watch closely next time you see films of Winston Churchill giving the "V-for-Victory" sign. Check to see who the spectators are. Often they are Americans, and the back of his hand is toward them.

The biggest hand-gesture problem for Americans is the "OK" sign, thumb touching forefinger in a circle. This is rude and obscene in many other countries of the world. (It refers to a circular part of the human anatomy.) I once saw a hilarious film montage of presidents of the United States going back to Herbert Hoover arriving at or departing from foreign countries and unintentionally insulting huge crowds of people with the OK sign. To my knowledge, the "thumbs up" sign is all right nearly everywhere and is a better choice. Two exceptions are Iran, where it is considered obscene, and Spain, where I just discovered it means "Long live the Basques!"

### Know Whom to Talk to

Beware of focusing your attention and communication on the most fluent English speaker. He or she may not be the most important person in the meeting, and you may be creating an offense. Be sure to have checked the business cards of people in the meeting and to have had some preliminary discussion as to the roles of the various people.

## USING INTERPRETERS

I speak and understand basic Japanese. I would not always mention that when I was in meetings with Japanese suppliers and without any Japanese co-workers. I understood enough to realize quickly that the supplier's interpreters were not always passing on what his team members were saying. Frequently, frank admissions of problems in Japanese were softened or even missing when put into English.

You should bring your own interpreter to any important meetings. At the very least, this will reduce the chances of the other party having an in-the-room caucus, which multiplies the supplier's effectiveness while a negotiation is going on. Your interpreter can be someone from your own company's offshore office. If you have a procurement office in that country, the procurement office should provide an interpreter. If your company has no office in that country, the U.S. Chamber of Commerce or the U.S. Embassy in that country can provide a list of interpreters. Be sure to specify that you need a business interpreter who understands technical terms.

---

### *Where Did They Learn to Speak English, Anyway?*

If the answer is England, any European country, or a former British colony, be alert for differences in the meaning of some apparently familiar words. There are many words naming common objects that vary between British and American English. A wardrobe is where you hang your clothes, not your collection of clothes. You put petrol in your car via the filler cap that is near the boot. You erase letters from a page with a rubber. The list goes on.

Fortunately, there are only a few business terms that seem to cause confusion. The biggest is "turnover," which the British use to refer to sales and Americans usually use to refer to departing employees. If you are told that the supplier's turnover is up 30 percent this year, don't conclude that there is a personnel problem. Another confusing term is "managing director," which means company president in the United States. If you are dealing in big numbers, be careful about anything over a billion. In England, a billion used to be a million million, in the United States it has always been a thousand million. The English used "milliard" for what we call a billion, and while most people have bowed to the Americanism, a few people still use "milliard." Be sure to check for understanding.

The real brand of English, as spoken in England, is taught in Europe and in former British colonies such as India and Singapore. American English prevails in most of the rest of the world. You'll be taking elevators and not lifts in Japan.

And, of course, remember that in England, WE are the ones with the accent.

If you do bring an unfamiliar interpreter, be absolutely sure to meet him or her at least a day ahead of the meeting so that the interpreter can become familiar with your key issues and any technical terms you might be using.

When you are using an interpreter, talk in short "sound bites" of approximately two sentences and allow the interpreter to translate. Again, consider appointing a "speed cop" from your team to be sure that you do not talk for long periods of time. Do not focus your attention on the interpreter, but concentrate on the person you are talking to. Don't ignore the interpreter either.

## CHECK FOR UNDERSTANDING

Throughout the meeting, someone should be writing down important points, conclusions, and agreements. This should be done legibly and visibly on a flip chart or electronic white board, so that the other side can see it. This is an important test that there is true understanding. If there has been a misunderstanding, it will likely be brought out into the open then. Volunteer for this job—power comes with the pen. Write down the conclusion as you best believe it, in simple, nondevious language. Your interpretation of the conclusion will likely prevail.

## DOES THE SUPPLIER HAVE TO KNOW ENGLISH?

In most circumstances, I believe the supplier must be able to communicate in English, at least in writing. The only circumstance where it is not necessary is if your company is buying a standard commercial product and your company has employees who speak the supplier's language.

If you are buying a product specifically designed for your company, your engineers must be able to communicate directly with the supplier's engineers, without any intermediaries. Technical communication is too complicated and too essential to require translation. If you have engineers that speak the supplier's language, you will also be OK, as long as your engineers stay with your company.

### What Can You Reasonably Expect?

How much should you expect a supplier to accommodate you by working in English? If a company is serious about selling internationally, it is reasonable to expect sales-oriented literature to be available in English. This would include product specifications, application instructions, and the like.

However, I don't think it is reasonable to expect documents that are intended largely for internal use to be translated. Quality manuals, personnel manuals, and manufacturing instructions are written for the seller's internal use. If you feel you need to examine them closely, get a copy in the supplier's language and hire a translator. In many cases you will have employees in your company who can translate the documents.

## *KEY POINTS*

▶ If a supplier is using English as a second language, the buyer should be responsible for preventing communication problems.

▶ To aid in communication, slow down, use more communication graphics, and eliminate sports and military metaphors from your language.

▶ Bring an interpreter to all but the most informal meetings. Allow an extra day to educate interpreters on your issues and vocabulary.

▶ Document, in writing, the conclusions and decisions made in a meeting prior to leaving.

## *RESOURCES AND REFERENCES*

Axtell, Roger. *Do's and Taboos of Body Language around the World.* New York: John Wiley and Sons, 1991. Somewhat repetitive, but good reading.

Bryson, Bill. *The Mother Tongue.* New York: William Morrow and Company, 1990. A lighthearted explanation of the complexities of the English language. Chapter 12, "English as a World Language," gives some insight into nonnative English speakers' use and misuse of English.

*Chapter Eight*

# Law

It's no secret that the legal system of the United States is in disrepute. The surprise is that our legal system was once the world's shining model. Over the last 50 years, that reputation has changed. Our system now has a reputation for litigiousness and so much attention to legal detail that the big picture can be lost. Legal habits and practices that U.S. supply managers use domestically will cause problems out of the country. You will have to adjust your legal practices to deal internationally. Fortunately, you can usually do that without compromising your real needs.

## U.S. LEGAL PRACTICES

The United States has some legal practices that most other countries do not. The biggest difference is that our legal system evolved from English common law, and most other countries' systems did not. Besides the difference in the legal system, we have some contracting practices that are unique to our culture. This chapter will explain those differences and give steps to take to avoid unpleasant surprises.

## OTHER LEGAL SYSTEMS

Most major industrial countries have equitable legal systems. It's possible to have a fair contract dispute resolution under any system of law that treats foreigners the same as it treats its own citizens. A global buyer should not be too concerned if a contract is written under the laws of Germany or Japan, for example.

Other countries, such as China, that are emerging into the world of international trade have much more questionable legal systems, which should be avoided. The general problem areas are the former Communist countries and third-world countries that are starting the development process.

Modern industrial countries have legal systems that came from one of two roots: common (or case) law or civil (or code) law.

### Case Law

This legal system originated in England and has been passed on to most former English colonies and possessions. U.S. law started on this basis (except in Louisiana, a former French colony). In case law, nearly all elements of a commercial contract are negotiable and must be spelled out. Precedent affects legal outcomes. Courts make judgments to cover situations that the written laws do not cover.

### Code Law

Code law is based on continental European legal theories. Romans originated it, and latter-day German and French legal scholars refined and renewed it. There are detailed, written codes, which are intended to cover all circumstances. Precedent is much less important. While judges still make judgments in difficult situations, this is regarded as unusual and a fault in the original codes.

Most of the non-English-speaking world uses code law. As countries emerged and modernized, as Japan did in the late nineteenth century, they generally chose code law over common law.

## CONTRACTS

Contracts are the key part of purchasing law. They are so ingrained in U.S. purchasing practice that buyers rarely question their need or purpose.

I can think of three reasons to have a contract:

1. A contract is a written record of the understandings between two companies.
2. A contract establishes buyer's rights and the new owners' obligations if there is a change of ownership.
3. A contract makes it more likely that both parties will perform according to plan, by improving the chance of a successful lawsuit.

You can accomplish the first two goals easily without overly compli-
cating the relationship. Much time and effort go into the third goal, and
in my opinion it is largely wasted. You will never need to sue a good
supplier. A bad supplier will likely be bad regardless of a contract. The
time it takes to pursue and achieve a legal remedy is much longer than
most buyers can tolerate.

Most companies will need to modify their legal practices when buying
offshore. They are likely to be dealing with a completely different legal
system.

### Code and Common Law Contracts

The contracting process is much simpler in code law countries because
negotiators can focus on the critical few needs. A typical American contract "default" clause is included in this chapter. I deliberately put it in
smaller print because I don't think anyone reads these clauses carefully.

### Default

If Seller breaches any provision of this agreement, Buyer may, by notice to Seller
and except as otherwise prohibited by applicable laws, terminate the whole or any
part of this agreement or any release, unless Seller:

> Gives Buyer notice, within three work days of receipt of Buyer's notice
> of its intent to cure the breach; and

> Cures the breach within 15 work days after receipt of Buyer's notice.

For purposes of Section 1 above, the term "breach" shall include without limitation
any:

> Proceeding, whether voluntary or involuntary, in bankruptcy or
> insolvency by or against Seller;

> Appointment, with or without Seller's consent of a receiver or an
> assignee for the benefit of creditors;

> Act by seller that endangers performance of this Agreement in
> accordance with its terms;

> Failure by Seller to make a delivery of products in accordance with the
> requirements of this agreement or any release;

> Failure to provide Buyer with reasonable assurance of future
> performance;

> Persistent or recurring failure to replace or rework products in a timely
> manner as required by Section 7.2 of this agreement; or

> Other failure to comply with the provisions of this agreement.

> In the event Buyer terminates this agreement in whole or in part as provided in section 1 above, buyer may procure, upon terms and in such manner as Buyer reasonably deems appropriate, products similar to the products as to which this agreement is terminated. Seller shall reimburse buyer on demand for all additional costs incurred by Buyer in purchasing such similar products. Seller shall continue the performance of this agreement to the extent not terminated under the provisions of this article.
>
> The rights and remedies granted to Buyer pursuant to Article 7 and this article are in addition to, and shall not be deemed to limit or affect, any other rights and remedies available at law or in equity.

The clause is extremely detailed. It has approximately 20 negotiable items, such as the number of days that parties have to respond. It is also extremely one-sided and will lead to arguments and discussions with any supplier who reads it and takes it seriously. A seller can be in default for one late delivery, for example.

In contrast, the same company's German subsidiary's standard purchase contract handles "default" in nine words: "In the event of default, German law will apply." Unless there is discomfort with the German definition of default and the remedies available to the buyer and seller, there will be no time spent in discussions and arguments about the default clause. Negotiators working under German law can move on to more important issues.

## Types of Agreements

You definitely need to have a written agreement with your supplier, foreign or domestic. However, the agreement does not have to look like a contract or even be called a contract. Both parties should sign it and both should expect to follow the terms of the agreement. However, cultural issues will affect both the form and the content of the agreement.

Suppliers in high-context societies (such as Japan) will tend to see the contract as a general guideline that sets the tone of a relationship. There is a general belief that the contract cannot cover every detail, and when the unexpected arises, each party will respect the other and try to accommodate the other party's legitimate needs.

Societies that are more legalistic, especially the formerly Communist countries, will have much more detailed, rigid contracts that are expected to be rigorously followed. However, there are entrepreneurial individuals who have actively rejected such formality. There will be a difference between state-owned companies and privately held companies.

American companies' attitudes to contracts vary widely. Some absolutely require them, but others form contracts only when pressured by suppliers. The companies that find themselves in U.S. courts frequently, using their contract to sue a domestic supplier for damages or to compel performance, are likely to have trouble with international contracts. Those that sign an agreement and put it in a file drawer where it is never seen again will have fewer problems.

If you do decide to have a formal contract with a foreign supplier, there are three possibilities for the formulation of the contract.

**U.S. domestic contract.**   You can have a U.S. domestic contract between your company and the U.S. subsidiary of the foreign company. This will probably be effective only if you buy from that subsidiary because otherwise there may be issues over the subsidiary's right to bind a parent. Your normal contract procedures will suffice.

**Foreign domestic contract.**   Your subsidiary can sign a contract in the supplier's country. This will probably be effective only if the subsidiary purchases the goods, as the same issues of the rights of the subsidiary to bind the parent will arise. This is the best alternative if your company has international procurement offices, which can provide major legal assistance.

**True international contract.**   This is a contract between a buyer and a seller who are in different countries. You need to take special care in negotiating and formulating these contracts.

## *Negotiating International Contracts*

If you do plan to negotiate a true international contract, you will need to prepare carefully. Some of the considerations in this section will also apply if you plan to have a domestic contract in a foreign country. The first issue is to decide what will be in a contract and what will be left out.

**Weed the contract mercilessly.**   Most U.S. companies have a set of contract "boilerplate" clauses that they have paid a lawyer handsomely to write. Generally, in the United States the clauses are complex and one-sided, like the "Default" clause shown earlier. In normal U.S. business practice, the buyer attaches the contract boilerplate to more specific requirements and gives the contract to the supplier. The supplier either signs quickly, objects and starts negotiating, or simply and quietly ignores the document. There is no real expectation that a seller will accept the boilerplate in its entirety.

If there is no signed contract, the equally one-sided boilerplate on the supplier's acknowledgment forms is the "last word." If there are differences, the UCC (Uniform Commercial Code) or normal commercial practice will apply. Business often goes on quite well in this fashion, because the terms are widely ignored.

However, if you pass this contract boilerplate to a foreign supplier who is unfamiliar with U.S. practice, the results can be unpredictable. There are schools of conflict resolution at the Harvard and Stanford law schools. Both of these have formulated a theory that an opening offer that appears unreasonable to the other side can prevent reaching an agreement even when there is a range of "win-win" possibilities. This is a scholastic confirmation of what seems to me to be common sense: Don't start a negotiation between buyer and seller by insulting the other party's intelligence or integrity.

You should remove clauses that are applicable only in the United States. Equal opportunity clauses are a good example. They will have no effect offshore. I support equal opportunity completely, but I see no reason to expect a German or Malaysian supplier to read about it in legal language. You should consult your own lawyer if you have any doubts as to the necessity of a clause.

You should also tone down any clause that seems too one-sided. Usually, it is easy to make the clause more reciprocal.

**Define laws and location.**    You will need to define the country whose laws will apply if a dispute arises that requires court action. If your company has a subsidiary in the supplier's country, I usually would agree to have the laws of the supplier's country apply, and the courts of that country administer the law. However, you should discuss this with your own attorney.

The clause defining the courts and laws needs more careful wording now that the United States has signed the Convention on the International Sale of Goods, the CISG.

The CISG is an international treaty, written by the UN. (It will be discussed later.) It applies to any international transaction where the buyer and the seller of goods (not services) are in countries that have signed the treaty, unless the buyer and seller specifically exclude it. It is not enough to say, for example, that the laws of California will apply. The CISG is a treaty signed by the United States and therefore attorneys can make a case that it applies in California. To exclude the CISG, you must specifically say that it does not apply.

**Consider an arbitration clause.**   If you really want to avoid courts, consider a clause that requires disputes to be settled by arbitration. The International Chamber of Commerce is a good source of arbitrators and has a good set of arbitration rules. This is another issue where you should seek general guidelines from your own attorney. Arbitration is fast and inexpensive, but often the decision is more of a compromise than a court would reach. You should also state in your contract that decisions by an arbitrator may be entered into a court for collection and enforcement.

---

*Sample Arbitration Clause*

A dispute between the parties that arises from this contract and which can not be resolved by the parties shall be settled by binding arbitration in accordance with the Rules of Conciliation and Arbitration of the International Chamber of Commerce, with three arbitrators, in Tokyo. Judgments of the arbitrators may be entered in any court which has jurisdiction.

---

## *Attitude toward Contracts*

In the Western world, we consider a contract negotiation to be over once the contract is signed. If there is a problem with the contract later, we might reopen negotiations, but we will do it reluctantly and with knowledge that the party reopening the negotiation is dealing from weakness.

In Asia, contracts are generally seen as more flexible. Because they cannot possibly cover every contingency, it is more normal to request a change during the life of the contract. This is not necessarily bad. It is usually done in good faith, with proper respect for the interests of the other party. Buyers can take advantage of it as well as sellers.

In 1995, there was an example of this practice being handled badly. McDonald's Corporation had a 20-year lease on a prime corner in Beijing, when suddenly the Chinese government decided they wanted the property. They unceremoniously booted McDonald's from their site. This is not the normal way of reopening contract negotiations and was another sign that the Chinese government had not fully accepted international business practices.

## CISG

The CISG is somewhat like the Uniform Commercial Code (UCC) in the United States, in both purpose and structure. They are both intended to fill in the gaps so that a contract need not be too detailed. However, there are significant differences in the formulation of contracts. One of the biggest differences is that large contracts under the CISG may be formed orally, while the UCC requires written contracts for any significant amount.

An additional difference is that the CISG is not so clear that a buyer has the absolute right to return goods for not meeting specifications. There is an obligation of the buyer to be sure the goods will not work in application, and to offer the seller the right to repair the damages in the buyer's country. The theory behind these items is to reduce shipping costs in international trade to the unavoidable minimum. In practice, it is often cheaper to have faulty goods repaired (at the seller's expense) in the buyer's country, rather than returning them.

---

*CISG Signers*

The following countries had signed the CISG as of late 1994. This list is a good test of the new world geography, as there are countries on it that just came into existence and whose names may not be familiar.

Argentina, Australia, Austria, Belarus, Bulgaria, Chile, Canada, China, Czech Republic, Denmark, Egypt, Finland, France, Germany, Hungary, Iraq, Italy, Lesotho, Mexico, Netherlands, Norway, Poland, Romania, Singapore, Slovakia, Spain, Sweden, Switzerland, Syria, Ukraine, U.S.A., Venezuela, Zambia.

The Treaty Section of the UN has an up-to-date list of signers and the text of the CISG. They can be reached in New York.

---

The CISG has some potential to be a "neutral" reference point if negotiating impasses arise. You can use specific individual clauses. However, most big companies have not chosen to rely on the CISG in its entirety because they do not seem to fully understand it yet.

## LEGAL PRIORITIES

Unless you foresee legal action in the future and plan to buy anyway, I recommend that you do not let legal considerations overrule good business judgment. Business often seems to get done well without any signed contracts. However, you do need some type of written document listing the expectations you have of the supplier and documenting the supplier's agreement.

The emphasis in the contract or other signed document should be on performance expectations, not on the penalties for not meeting them and not on a detailed plan for something everyone hopes will not happen. You can write performance expectations for on-time shipment rates, price competitiveness, product, quality or lead-time reduction. You can write these in simple English, without the involvement of lawyers. They are really the essential part of an agreement.

## INTELLECTUAL PROPERTY

One of the more troublesome aspects of buying internationally is the difficulty in enforcing copyright, patent, and trademark rights. This usually shows up as a problem when a buyer discloses design information to a potential (or actual) supplier, who copies the product and resells it at a lower cost. The supplier has received free research and development.

The chances of this happening are a strong function of the country that the seller is in. Some countries take a very casual attitude toward intellectual property registrations of other countries. There are four key international treaties and conventions that a country may have agreed to. Agreement will give some assurance that the country is likely to give respect to intellectual property rights. This information is in *International Business Practices,* a convenient reference work that is described in the resources section of this chapter. The four key treaties are as follows:

## WIPO

The World Intellectual Property Organization is a Geneva-based UN organization that promotes and administers treaties between individual countries.

### Paris Convention

The Paris Convention (formally, The International Convention for the Protection of Industrial Property) sets some standards for patents and trademarks, as well as for the applicability of one country's trademarks in another country.

### Berne Convention

The Berne Convention for the Protection of Literary and Artistic Works gives minimum standards for copyright protection.

## UCC

This UCC is the Universal Copyright Convention, and is similar to the Berne Convention.

*International Business Practices* briefly describes many countries' intellectual property practices.

### Practical Test

The number of treaties a country has signed is only one indication of the attitude toward intellectual property. China, where software pirating is common, has signed all four treaties, for example. My favorite practical test is to check hotel bookstores and local software prices. If you find hardcover English-language books, or local-language versions of major commercial software, at a fraction of the U.S. price, that is a sign that intellectual property protection is poor. So is the open selling of fake famous-label watches and clothing.

### Protecting Yourself

In countries where there is a problem, you should be sure to understand that the legal system will give no practical protection. Your major protection is to choose the right supplier as your partner. If you're just getting started with a supplier, you will have to do some very thorough reference checks on the supplier and its management. This means the supplier must have engaged in international sales for some time. You should not disclose crucial information to a potential supplier who does not have a good history.

## FOREIGN CORRUPT PRACTICES ACT

You should be aware that you can be held criminally accountable under U.S. law for bribery or other acts you commit in a foreign country, even if the act is legal in that country. I am not qualified to advise on this criminal law matter, but if you find yourself in a situation where you anticipate a request for a bribe, seek legal advice from both a U.S. and a foreign attorney before proceeding. In some cases a "payment to expedite service" to a lower-level government employee may not be the same as a bribe.

Mexico, Korea, and some other countries have a reputation for being very prone to small-scale bribery requests by local officials. You need to have a solid company policy on this matter. Once you make the first payment, you will probably get requests for more. A firm refusal at the start may lead to some short-term delays but can have benefits in the future. I recommend allowing some extra time for the first few export transactions from these countries so that you are not pressed into making unfortunate payments.

## RECIPROCITY

In the United States and Canada, it is illegal under antitrust law for a company's customer to demand that the seller reciprocate and buy material from the customer. In other countries it is quite common and not illegal. Do not be surprised if your company's foreign customers apply some pressure on you to buy from them based on their purchases from you.

You should consider it open-mindedly, and if buying is at all feasible, allow them to make a sales presentation. This will help your company's sales efforts. However, I don't recommend relaxing purchasing standards at all for sales purposes.

## UNIONS

Many U.S. companies are hesitant to buy from unionized companies. Unions are different overseas. Many countries have laws that require all manufacturing companies to have unions representing the employees and to consult the employees before making major changes. Japan and Germany are examples. In Germany, employee representatives must be on the board of directors of manufacturing corporations.

If you are concerned about unionization, first check whether unions are required by law. If they are required, then check how long it has been since a work stoppage or strike. Many unions work closely with management in a mutually beneficial relationship.

## *KEY POINTS*

▶ The United States uses common or case law, which results in lengthier and more detailed contracts than are found in countries that use code law or civil law.

▶ There should be a written and signed document that describes the expectations of buyer and seller. It does not have to look like a U.S. contract.

▶ Contracts with foreign suppliers can be domestic in either country if both parties have a legal presence there. They can also be true international contracts. In this case, exclude the CISG, and do not count on the legal system to provide timely relief.

▶ Advanced, industrial countries have legal systems that can be trusted to treat foreign companies fairly. Developing countries may not.

▶ There is no effective legal protection in many countries against intellectual property piracy. The only protection is through a thorough reference check of prospective suppliers.

## RESOURCES AND REFERENCES

Folsom, Ralph; Michael Gordon; and John Spanogle. *International Business Transactions in a Nutshell.* St. Paul, MN: West Publishing Company, 1992. This is a book for lawyers who are getting started in international trade transactions. Non-lawyers can understand it too.

U.S. Department of Commerce, *International Business Practices,* is available from the U.S. Government bookstores. It is aimed at helping exporters from the United States, but has useful information on every country.

King, Donald, and James Ritterskamp. *Purchasing Manager's Desk Book of Purchasing Law,* 2nd ed. Englewood Cliffs, NJ: Prentice Hall, 1993. This has a very thorough analysis of the CISG.

## Chapter Nine

# Introduction to Currency

This chapter is the first of seven chapters on foreign exchange. It covers some of the simpler aspects, such as how to convert prices from one currency to another, the definition of weaker and stronger currencies, and the concept of exchange risk. There is a test of understanding at the chapter's end. Those who are familiar with these topics may prefer to skip this chapter.

## THE CURRENCY PROBLEM

With all the increased ease of international commerce, the necessity to exchange currencies remains one of the most intractable problems. Why is there such an attachment to having separate national currencies? It's not just nationalistic pride.

Having control of a currency gives a country a very important fiscal tool: the control of interest rates. A government can raise rates to stifle inflation or lower rates to ward off a recession. If more than one country uses the same currency, it will be impossible to maintain different interest rates in the different countries. Borrowers would borrow in the lowest-rate country and transfer the money to where they needed it.

### *Attempts to Stabilize Currencies*

Currently, most major currencies "float" against each other. This is a relatively recent development. Up until 1973, there was a worldwide attempt to maintain fixed exchange rates. In Europe, a second attempt to stabilize European currencies against each other took place from the late 1980s until late 1992, when it largely collapsed.

**Bretton Woods Agreement.** In 1944, the countries that expected to win World War II set up an exchange rate control mechanism that lasted until 1973. Under this agreement, called the Bretton Woods Agreement, the U.S. government guaranteed to sell an ounce of gold for $35. All other countries guaranteed to keep their currencies at a fixed value versus the dollar by raising and lowering interest rates and by controlling imports. A higher (real) interest rate drives up the value of a country's currency, as it makes investment in that country more attractive. Reduced imports also drive up the value of a currency by reducing pressures to sell that currency for the currency needed to pay a foreign supplier.

This mechanism lasted until 1973. During this period, rates did occasionally change, but only after a great deal of discussion between the involved countries' finance ministries. The mechanism left the problem of exchanging currencies intact but prevented today's uncertainty of the day-to-day exchange rates between currencies.

The agreement collapsed in 1973, when the United States eliminated convertibility of the dollar into gold. Part of the reason was that the United States was experiencing inflation, and part of the reason was that in 1971 the United States developed a negative balance of trade for the first time since World War II. Also during that time, the application of electronics to currency trading began to develop, and the ability of any government to control the value of its currency began to decline. It became so easy to transfer funds electronically between private parties that the transaction volume became larger than the capabilities of governments to influence prices.

**European attempts at currency unity.** The floating rates that arose after 1973 caused a lot of confusion in international trade. In the 1980s, the European Union (then called the European Economic Community) attempted to increase stability in intra-European trade by fixing the exchange rates of each of their member countries against the others. In most cases, exchange rates were allowed to fluctuate plus or minus 2.5 percent against a nominal value of other European currencies. The European currencies as a group could float freely against non-European currencies.

This arrangement came to be known as the "snake," as various currencies zigzagged across a narrow band, which itself moved up and down against the dollar. The pattern on a graph looked snakelike.

This arrangement was intended to be the first in a series of steps moving to the use of one currency in Europe. However, it became painful to a few countries, especially Britain and Italy. Their currencies were tending to get weaker due to inflation and a poor balance of trade, and their governments were forced to raise interest rates to strengthen the currencies. The high interest rates drove the countries toward recession, and domestic politics came to the front.

In September 1992, Britain and Italy dropped out of the mechanism. The British pound dropped 12 percent against the U.S. dollar and other European currencies in a matter of hours. (I was flying home that day. When I left my British office in the morning it took nearly two dollars to buy a British pound. When I arrived in California, it took only $1.75.)

Currently, major industrial countries' currencies float freely against each other, with the exception of the core European countries of Germany, France, and the Benelux countries of Belgium, the Netherlands, and Luxembourg. While these currencies could float, present attempts to maintain a stable (±2.5 percent) exchange rate (among themselves, not against the U.S. dollar or Japanese yen) are successful.

This is where we are today. We have to exchange currencies and convert foreign currency prices into dollars to make good sourcing decisions.

## CONVERTING PRICES

The formulas for converting currency are simple.

To convert prices from one currency to another, you use the formula

$$\text{Dollars} = \frac{\text{Foreign Currency}}{\text{Exchange Rate}}$$

This can also be written as

$$\text{Foreign Currency} = \text{Dollars} \times \text{Exchange Rate}$$

You can find the exchange rate tables in any daily financial paper, such as *The Wall Street Journal*. Excerpts from a currency table are shown in Table 9–1. The exchange rate to use is in the column titled "Currency per U.S. Dollar." To remember this, remember that the rate should always be a number larger than 1, except for British and Irish pounds and the Brazilian real (and a few other minor currencies that most buyers are unlikely to encounter, such as the Maltese lira). Only a few currencies are worth more than one U.S. dollar.

**TABLE 9–1**
*Currency Values, Friday June 30, 1995 (selected countries)*

|  | U.S. Dollar Equivalent | | Currency per U.S. Dollar | |
| --- | --- | --- | --- | --- |
| Country | Friday | Thurs. | Friday | Thurs. |
| **Britain** (pound) | 1.5945 | 1.5990 | .6272 | .6254 |
| 30-day forward | 1.5937 | 1.5970 | .6275 | .6262 |
| 90-day forward | 1.5913 | 1.5947 | .6284 | .6271 |
| 180-day forward | 1.5858 | 1.5895 | .6306 | .6291 |
| **Germany** (mark) | .7236 | .7247 | 1.3820 | 1.3798 |
| 30-day forward | .7247 | .7256 | 1.3799 | 1.3782 |
| 90-day forward | .7263 | .7272 | 1.3768 | 1.3750 |
| 180-day forward | .7284 | .7296 | 1.3729 | 1.3707 |
| **Japan** (yen) | .01180 | .01184 | 84.710 | 84.480 |
| 30-day forward | .01187 | .01187 | 84.279 | 84.224 |
| 90-day forward | .01196 | .01197 | 83.637 | 83.557 |
| 180-day forward | .01210 | .01212 | 82.615 | 82.520 |
| **Mexico** (new peso) | .1603 | .1599 | 6.2400 | 6.2550 |
| **Taiwan** (new Taiwan dollar) | .03871 | .03873 | 25.828 | 25.818 |

Table 9–1 shows more than one rate for several currencies. Some of the rates are labeled "forward." These forward rates will be explained later. Use the first rate listed, and be sure to use the column for the right day. In this example, the correct column is the Friday column. The exchange rate for the German mark is 1.3820 marks per U.S. dollar. The exchange rate for the Japanese yen is 84.710 yen per dollar.

How many dollars are 240 Japanese yen (¥) worth at these exchange rates? Dollars are equal to foreign currency divided by the exchange rate: ¥240 divided by 84.710 yen per dollar is $2.8332.

How much is $4.50 in German marks (also known as deutsche marks, abbreviated DM)? Foreign currency equals dollars times the exchange rate: $4.50 times 1.3820 marks per dollar is DM6.2190.

## WEAKER AND STRONGER

Currencies strengthen and weaken with respect to each other. There always has to be a reference point. We can say, "The dollar weakened against the German mark," or we can say, "The German mark strengthened against the dollar." These mean exactly the same thing.

The dollar is weakening against a foreign currency when the exchange rate goes down. The exchange rate for the Mexican peso moved from 6.2550 pesos per dollar on Thursday to 6.2400 pesos per dollar on Friday. You get 0.015 fewer pesos per dollar on Friday than you could get on Thursday.

Conversely, you can get more yen per dollar on Friday than you could on Thursday. The dollar is worth 0.230 (84.710 minus 84.480) more yen on Friday. This means the dollar strengthened against the yen.

You could also say that the yen weakened against the dollar by 0.230 yen. However, it's easier to keep your terminology straight if you focus on the dollar and remember, "Weaker dollar means lower exchange rate, stronger dollar means higher exchange rate." If someone talks of the yen weakening against the dollar, mentally change the words to reflect that the dollar is strengthening against the yen. The exchange rate will be increasing.

The impact of a weaker dollar on buyers is that imported goods get more expensive. Consider a product that a Japanese manufacturer will sell for 200 yen (¥200). If the exchange rate is 85 yen per dollar, this part will cost an American $2.3529. (Dollars equal foreign currency divided by exchange rate: ¥200 divided by 85 yen per dollar equals $2.3529.) However, if the dollar weakens to ¥80 per dollar, and the Japanese manufacturer still sells the product for ¥200, it will cost an American more than before. It will cost $2.50. (Two hundred yen divided by 80 yen per dollar equals $2.50.)

If the dollar strengthens, the opposite happens. The ¥200 product will cost an American less than the original $2.3529. If the exchange rate were to move from ¥85 per dollar to ¥90 per dollar, the ¥200 part would now cost an American approximately $2.22. (Two hundred yen divided by 90 equals $2.2222.)

## EXCHANGE RISK

If the price of a product is set in one currency, the user of the other currency is taking on the "exchange risk." If an American buyer is buying a product whose price is set in yen, the American buyer is taking on the exchange risk. Conversely, if a Japanese seller sets a price in dollars, the seller is taking on the exchange risk. There is a risk that the exchange rate could change between the time the price is set and payment is made. A rate change will benefit one party and hurt the other.

Here is an example of how this works:

Assume today's exchange rate is 85 Japanese yen per dollar. You want to order a product for delivery in six months that costs 10,000,000 Japanese yen. Today that is equivalent to $117,647. (Foreign currency divided by exchange rate equals dollars: ¥10,000,000 divided by 85 yen per dollar equals $117,647.)

## Supplier Accepts Exchange Risks

If the supplier sets the sales price in dollars, he or she is taking on the exchange risk. Six months from today, when the supplier sends an invoice for $117,647, the exchange rate of the yen could have moved up or down as much as 25 percent. To take a moderate example, assume the rate has changed plus or minus 10 yen, which is an 11.8 percent change.

**Weaker dollar.** If the exchange rate has changed 10 yen downward (e.g., from ¥85 per dollar to ¥75 per dollar) the dollar has weakened against the yen.

The $117,647 is now worth ¥8,823,525. (Dollars times exchange rate equals foreign currency: $117,647 times 75 yen per dollar equals ¥8,823,525.) You still pay $117,647, but the supplier is receiving ¥1,176,475 less than he planned. Your supplier will not be happy.

**Stronger dollar.** If the exchange rate has changed by 10 yen upward (from ¥85 to ¥95 per dollar) the dollar has strengthened against the yen.

The supplier still invoices you for $117,647, which is the agreed-upon price. The $117,647 is now worth ¥11,176,465. (Foreign currency equals dollars times exchange rate: $117,647 times 95 yen per dollar is ¥11,176,465.) The supplier is getting ¥1,176,645 more than he had anticipated. He's going to be quite happy, as he had an unexpected profit.

You are not unhappy unless you stop to think that you would have been better off to take on the exchange risk. If the supplier had billed you for ¥10,000,000, your cost would have been ¥10,000,000 divided by 95 yen per dollar, or $105,263, not $117,647.

Note that the term "exchange risk," although widely used, is not quite right. "Exchange uncertainty" is a better term. The party taking on the exchange risk might benefit or might have a loss.

### Buyer Takes Exchange Risk

Now, let's assume instead that you took the exchange risk by setting the price at ¥10,000,000 instead of $117,647.

**Weaker dollar.** If the dollar weakens by ¥10, the exchange rate will change from 85 yen per dollar to 75 yen per dollar. It will cost you $133,333 to buy the ¥10,00,000 to pay the supplier. (Foreign currency divided by exchange rate equals dollars: ¥10,000,000 divided by 75 yen per dollar equals $133,333.) The supplier gets his 10 million yen, exactly what he planned on, but you are not happy because you spent $15,866 more than you expected.

**Stronger dollar.** If the dollar strengthens by ¥10, the exchange rate will change from 85 yen per dollar to 95 yen per dollar. It will cost you only $105,263 to buy the ¥10,000,000 to pay the supplier. (Foreign currency divided by exchange rate equals dollars: ¥10,000,000 divided by 95 equals $105,263.) You are happy because you paid fewer dollars than you expected. The supplier got what he expected and is also happy.

## WHY TAKE THE RISK?

Why would a buyer agree to take on exchange risk? To get a lower initial price. In this example, there were two choices for the price: ¥10,000,000 or $117,647. They were worth exactly the same at the assumed exchange rate of 85 yen per dollar. However, in the real world, a buyer should be able to achieve a lower price for taking on the exchange risk. The yen price could have been ¥9,000,000 or ¥9,500,000. The initial dollar cost could have been $105,882 or $111,765.

Later chapters will describe when to set prices in dollars and when to set them in foreign currency. They will also describe how a buyer who takes on the exchange risk can protect his or her company against unexpected price changes.

*Test of Understanding*

Assume today's exchange rate against the German mark is DM1.70 per dollar. You are making a purchase with a value of DM170,000. You will make payment six months from now.

1. What is the dollar equivalent today?

2. If the dollar strengthens 5 percent against the mark, what will the exchange rate be?

3. If the dollar weakens 5 percent against the mark, what will the exchange rate be?

4. Assume the supplier takes the exchange risk. How many dollars will you pay and how many German marks will the supplier receive if the dollar weakens 5 percent from 1.70?

5. The supplier still takes the exchange risk. How many dollars will you pay and how many marks will the supplier receive if the dollar strengthens 5 percent from 1.70?

6. Now you take the exchange risk. How many dollars will you pay and how many German marks will the supplier receive if the dollar is 5 percent weaker?

7. You still have the exchange risk. How many dollars will you pay and how many German marks will the supplier receive for a 5 percent stronger dollar?

Answers are in Appendix A at the end of the book.

## KEY POINTS

▶ Foreign currency divided by exchange rate equals dollars.

▶ Dollars multiplied by exchange rate equals foreign currency.

▶ If a currency's exchange rate per dollar goes up, the dollar has strengthened against that currency. Stronger dollars make it cheaper for U.S. buyers to buy offshore,

▶ If a currency's exchange rate per dollar goes down, the dollar has weakened against that currency. Weaker dollars make it more expensive for a U.S. buyer to buy offshore,

▶ The party (buyer or seller) whose currency is not being used is taking on the exchange risk.

## *RESOURCES AND REFERENCES*

Weissweiller, Rudi. *How the Foreign Exchange Market Works.* New York: New York Institute of Finance, 1990. This is a short book that gives the basics of currency trading. It is a good reference for all the currency chapters.

*Chapter Ten*

# Currency Selection

Americans are generally unfamiliar with the workings of the foreign exchange markets. Unless we are on a holiday trip out of the country, most of us will rarely deal with a foreign currency. This is a result of both our geographic isolation and the large role that the U.S. dollar plays in the global economy. In contrast, almost every other country has foreign exchange desks in most branch banks, and independent companies that do nothing but exchange money. Changing money is a very common occurrence.

Partly because of unfamiliarity with other currencies, many buyers prefer to pay in dollars. They often believe that this is the most risk-free way to buy. They will occasionally see headlines describing major losses in currency markets and be even less inclined to work in foreign currencies. (These losses were usually due to speculation, which is not a practice I advocate.) Company treasury departments may also be reluctant to permit buying in a foreign currency. I believe that buying in dollars is a costly mistake in many cases. It often leads to higher-than-necessary prices and does not reduce risk appreciably.

Sometimes you should buy in U.S. dollars and other times in a foreign currency. The correct choice depends on two factors:

- The type of product you are buying.
- The country you are buying from.

## KEY DEFINITIONS

Before continuing, some words need defining.

### *Pricing Currency*

The pricing currency is the currency used to set the purchase price. If you set the price in the seller's currency, the buyer is taking on the exchange risks. The price might be set as "1.60 deutsche marks" (DM1.60) or "120 yen" (¥120). The supplier will want a set amount of his currency in payment. He does not particularly care how much that currency is worth in dollars. Contracts would specify the price in the seller's currency.

If you set the price in the buyer's currency, the seller is taking on the exchange risks, as was explained in Chapter 9.

## Payment Currency

In some cases the buyer may pay in a different currency than prices were set in. The supplier might receive the dollar equivalent of the foreign currency price, or perhaps the foreign currency equivalent of a dollar price.

This can occur in highly inflationary economies that import parts for assembly. For example, when Mexico had a high inflation problem, prices for assemblies made from foreign (U.S.-built) goods were often set in dollars. Mexican law requires payment in pesos for domestic Mexican purchases, so purchase orders were written with prices stated as "The peso equivalent of X.XX dollars."

The opposite can also happen. Korea does not permit payment for exports in the Korean currency (the won). However, prices can be set in won, and purchase orders can read "The U.S. dollar equivalent of XXX won." This approach can also be used in dealing with the U.S. domestic subsidiary of a foreign supplier. They might accept payment in dollars for imported parts with their price set in the manufacturer's currency.

The biggest savings will come from proper selection of the pricing currency. The actual currency paid usually matters very little. Either the buyer or seller can exchange one currency for another for a very small cost.

## Manufacturer's Currency

This is the currency in which a manufacturer incurs costs. It will normally be the local currency of the supplier's country. The supplier needs this currency to pay salaries, utilities, taxes, and building costs.

The parts and material that a supplier buys may or may not be priced in the currency of the supplier's country. If the supplier is an assembler in a small or developing country, he or she is very likely to be importing parts for manufacturing. These parts may be priced in a third currency, different from either the supplier's currency or the U.S. dollar. Assemblers in Taiwan, for example, buy a lot of Japanese parts, and sometimes the majority of the supplier's material costs will be in yen. It's very possible to have more than one "manufacturer's currency" in these circumstances.

However, to keep the discussion simple, assume for now that there is only one manufacturer's currency per product. A later chapter will cover more-complex situations.

## Floating Currencies

Floating currencies are the major, freely traded currencies. Free-market forces (with occasional government intervention) set exchange rates. Countries with floating currencies include the United States, Japan, Canada, Australia, New Zealand, and all the countries of western Europe. These countries' currencies can fluctuate against the dollar quickly and unpredictably. The value can change 10 percent a month for a number of months in a row.

## Pegged Currencies

These are currencies of countries whose governments attempt to maintain a fixed exchange rate against the U.S. dollar. Typically, these are smaller or developing countries. The degree of success in maintaining fixed rates varies from country to country. The Hong Kong dollar and the Thai baht have not moved more than a percent or two against the U.S. dollar in the last 10 years. The Taiwan and Singapore dollars have been relatively stable from day to day or week to week, but both have gone through extended periods of slow, steady, and controlled strengthening at one time or another.

**Pegged currency economics.** While a pegged currency is usually steady against the U.S. dollar, there are cases where the peg has not held under stress. The currency of Taiwan strengthened against the U.S. dollar in 1988, and the currency of Mexico weakened against the U.S. dollar in 1995.

Pegging works when a country with a small economy has (or can buy) enough dollars in reserve to be able to guarantee that they will give a fixed amount of dollars for their currency. They get those dollars by running a positive balance of trade with the United States, or with another hard-currency country whose currency can be traded for dollars. However, if they get too many dollars by running a too-positive balance of trade, then political factors arise that can cause their currency to strengthen (or the dollar to weaken against that currency).

There are two factors to consider when evaluating the ability of a country to maintain a currency pegged to the dollar:

- The country's balance of trade.
- The country's hard-currency reserves.

If a country is running a negative balance of trade, there will be pressure to devalue the currency. This happened in Mexico in 1995. Mexico had been trying to maintain a "sliding peg" of the peso against the dollar. The peso was being allowed to weaken a small amount every day. However, the surge of imports into Mexico that resulted from the North American Free Trade Agreement (NAFTA) caused Mexico's trade balance to turn strongly negative. For a while, they were able to compensate by using dollar reserves.

A high inflation rate in Mexico caused additional pressure. Prices were going up in pesos, but the peso was not devaluing enough to avoid having the dollar value of Mexican prices go up. This reduced Mexican competitiveness, and therefore reduced Mexican exports.

The two pressures combined to force the Mexican government to abandon attempts to peg the peso to the dollar. When they did so, the dollar strengthened against the peso from approximately 3.2 pesos per dollar to nearly 6 pesos per dollar in a matter of weeks. (This means the peso weakened against the dollar.) Exports from Mexico suddenly became cheaper in dollars.

In most cases, the reaction of a country facing a potential devaluation of a pegged currency due to poor balance of trade will be to take steps to reduce imports. However, because of NAFTA, the Mexican government was unable to restrict imports by duties or other indirect means, which left no alternative but devaluation.

In comparison, in 1986, Taiwan had a very positive balance of trade with the United States. They had been maintaining their currency (New Taiwan dollar, or $NT) at $NT42 per U.S. dollar for years. Their trading partners, particularly the United States, pressured them to strengthen their currency because their prices, when expressed in dollars, were too low. They agreed to do this in a controlled fashion, and their currency strengthened (another way of saying the dollar weakened) from 42 per U.S. dollar to approximately 28 per U.S. dollar. They did this by slowly changing the amount of New Taiwan dollars they would give or charge for each U.S. dollar.

Some pegs are more rigid than other pegs. The country's currency may vary a few percent either way from day to day or week to week, and some countries will vary more than others. Appendix C of this book shows the amount of some key currencies' fluctuation graphically.

## TYPE OF PRODUCT PURCHASED

The first factor to consider in selecting a currency is the type of product that you are buying. Generally, you can divide purchased products into two categories:

- Market-driven products
- Cost-driven products

You need a different currency strategy for each category of parts.

### *Market-Driven Products*

These are commodities that have multiple, easily changeable manufacturers or sources. Examples are gold, oil, raw metals, and some electronic components such as dynamic random access memories (DRAMs). Generally, the worldwide market price is in U.S. dollars. If a DRAM costs $10.00 in the United States, the price will be within a very small percentage of the equivalent amount of yen in Japan or deutsche marks in Germany. Manufacturing cost has very little to do with the sales price of these products. Prices are set by pure competition and controlled by the most expert negotiators and most efficient manufacturers.

When you buy these products, you should set prices in dollars, even if you make payment in another currency. There is no point in setting prices in a foreign currency. If you price in a foreign currency, the price will be much more variable than if you price in dollars because the foreign currency prices would have to vary to match the dollar price.

If the dollar prices change rapidly, as they do during the first few months of sale of a new generation DRAM, you need to negotiate new dollar pricing frequently.

### Cost-Driven Products

Other products are cost driven. You will recognize these because you will be having extensive discussions about the supplier's costs. You will probably be getting costed bills of materials from the supplier. These products are typically custom-designed and are built only by one supplier. They are difficult to transfer between suppliers. Mechanical parts, computer monitors, consumer appliances, and power supplies are cost driven.

## MANUFACTURER'S COUNTRY

The other factor in choosing a currency is whether the manufacturer's currency is pegged to the dollar or floating.

### Pegged Currency

If you are buying from a country with a currency pegged to the U.S. dollar, the supplier's currency is, for practical purposes, the U.S. dollar. You can price either in dollars or the supplier's currency. This applies to both market-driven and cost-driven products.

Because the currency is pegged to the dollar, there is very little exchange risk. If the supplier will permit dollar pricing, choose dollars because this is the easier choice. The supplier may want an "escape clause" (see Chapter 15) to protect against a dollar devaluation if you price in dollars, and you should have one if you price in the supplier's currency.

If the supplier from a pegged-currency country is assembling a cost-driven product with materials from other countries, it might be necessary to protect against movement of those currencies. This will be covered in the section on multiple-currency products.

### Floating Currency

Floating-currency countries are the countries where the currency complications arise.

**Cost-driven products.**   You should price a cost-driven product from a floating-currency country in the supplier's currency. You should get a lower price for doing so because you are relieving the supplier of currency risk. Examples of these products are monitors from Japan and mechanical parts from Europe.

I believe that a price decrease of between 5 and 10 percent is usually achievable by pricing in the supplier's currency. You will be starting with the lowest price possible.

It is true that if the dollar weakens, the dollar cost will go up. You can delay this increase by hedging. It's also true that the dollar cost will go down if the dollar strengthens.

In comparison, if you price in dollars, the dollar cost will usually still go up if the dollar weakens significantly. I have seen too many honorable, well-intentioned suppliers forced to raise their supposedly fixed dollar prices when the dollar weakens. However, I have yet to see a supplier volunteer to lower his prices when the dollar strengthens. Pricing in dollars gives only the illusion of security.

**Market-driven products.** You should price a market-driven product from a floating currency country in the market currency, which is almost always U.S. dollars. If the product is priced in the supplier's currency, frequent renegotiation of the local currency price will be necessary as the exchange rate changes. Examples of these products are wheat from Canada and DRAMs from Europe.

**TABLE 10–1**
*Pricing Currency*

| | *Pegged-Currency Country* | *Floating-Currency Country* |
|---|---|---|
| **Cost-Driven Product** | Price in dollars or the manufacturer's currency | Price in the manufacturer's currency |
| **Market-Driven Product** | Price in dollars | Price in dollars |

## SUMMARY

Table 10–1 shows the best choices for pricing currencies. In all these cases, you will have to take steps to protect yourself against currency fluctuations. This can be through hedging, risk sharing, or escape clauses, all of which will be covered in later chapters.

## KEY POINTS

▶ Market-driven products are those whose price is set by short-term supply and demand. Often, the market price is in dollars, and the price is very similar anywhere in the world. These products can be priced in dollars without additional costs.

▶ Cost-driven products are those where the cost of manufacturing is a major determinant in setting prices. Prices for these products should be set in the supplier's currency at a lower initial price than if the prices were set in dollars.

▶ Some countries' currencies are pegged to the U.S. dollar. The dollar can be considered the supplier's currency in those cases, but an "escape clause" will be needed.

# Chapter Eleven

# Introduction to Hedging

Several times a year the financial press reports that a company has lost huge amounts in the foreign-exchange markets. In 1993, the entire top management of Showa Shell (Shell Oil's joint venture in Japan) resigned after an exchange loss of $1.05 billion. In every recent case, the major unexpected losses were the result of speculation, which is different from hedging.

Hedging is protecting the value of a particular asset, liability, or future cash flow. If your company sells to foreign customers in their own currency, it may already be hedging sales orders. By hedging the foreign currency value of an open sales order, the dollar value of that order can be locked in. Your company is then protected against exchange-rate changes between booking an order and receiving payment. The protection is specific to one cash flow and limited to the amount necessary to protect it.

Showa Shell was speculating. Their financial management felt that the dollar would strengthen to 150 yen per dollar. They did not try to protect a particular asset. Instead, they made commitments on the financial markets that would have made money if the dollar strengthened, but they lost money when it weakened. Even worse, when the managers first started losing money, they made additional investments so that they would not have to report a loss. Finally, they were more than a billion dollars behind and had to make a public announcement.

In most companies, purchasing is the first or second largest outward cash flow. (Salaries and wages are the other large flow.) If a significant portion of the purchasing expense is denominated in a foreign currency, this cash flow should be hedged.

Hedging is not totally risk-free, but you can estimate the risks ahead of time and control them. The risk comes from purchase volume forecasts that do not come true.

The chapters on hedging will proceed in four steps. First, this chapter will explain why exchange rates change and the effect of changes on your costs. Chapter 12 gives a simplified explanation of forward contracts, and Chapter 13 gives a simplified explanation of currency options. Chapter 14 clarifies the simplified explanations, explains the risks involved, and shows the right hedging decision.

## SPOT RATES

Almost every daily newspaper has listings of the exchange rates of most major countries. Please refer to Table 11–1, which is the same as Table 9–1, discussed earlier. In this table, some countries (Britain, Japan, and Germany) are listed with several exchange rates. Others (Mexico and Taiwan) show only one rate. Currency traders call the top line for Britain, Germany, and Japan the "spot rate." The one line for Mexico and the one line for Taiwan are also the spot rate. This is the rate for an immediate transaction. If you traded a dollar for German marks on Friday, June 30, 1995, you would have received 1.3820 marks, minus bank charges.

**TABLE 11–1**
*Currency Values, Friday, June 30, 1995 (selected countries)*

| Country | U.S. Dollar Equivalent | | Currency per U.S. Dollar | |
|---|---|---|---|---|
| | Friday | Thurs. | Friday | Thurs. |
| **Britain** (pound) | 1.5945 | 1.5990 | .6272 | .6254 |
| 30-day forward | 1.5937 | 1.5970 | .6275 | .6262 |
| 90-day forward | 1.5913 | 1.5947 | .6284 | .6271 |
| 180-day forward | 1.5858 | 1.5895 | .6306 | .6291 |
| **Germany** (mark) | .7236 | .7247 | 1.3820 | 1.3798 |
| 30-day forward | .7247 | .7256 | 1.3799 | 1.3782 |
| 90-day forward | .7263 | .7272 | 1.3768 | 1.3750 |
| 180-day forward | .7284 | .7296 | 1.3729 | 1.3707 |
| **Japan** (yen) | .01180 | .01184 | 84.710 | 84.480 |
| 30-day forward | .01187 | .01187 | 84.279 | 84.224 |
| 90-day forward | .01196 | .01197 | 83.637 | 83.557 |
| 180-day forward | .01210 | .01212 | 82.615 | 82.520 |
| **Mexico** (new peso) | .1603 | .1599 | 6.2400 | 6.2550 |
| **Taiwan** (new Taiwan dollar) | .03871 | .03873 | 25.828 | 25.818 |

## Market Forces Affecting Spot Rates

Until the early 1970s, the theory of exchange rates was that they were controlled by trade deficits and surpluses. The theory was that trade deficits would drive a country's currency down, and trade surpluses would raise the value of the currency. In the simple case of a two-country world, the country that was a net importer would have to sell its currency for the exporter's currency to be able to pay the exporter. This would drive down the value of the currency of the importer and raise the value of the exporter's currency. This in turn would make the net importer's products more competitive, and start to raise the export level. At some point, there would be an equilibrium in currency values and trade.

Like a lot of what I learned in economics classes, this is no longer true. Today, with electronic fund transfers and global currency markets, interest rates are the main driver of currency values. There is 20 or 30 times as much money transferred in the foreign-exchange markets as there is international trade in goods and services.

A corporate treasurer with some spare cash can move the cash to a short-term account in a foreign bank that offers higher interest rates, or can buy foreign bonds. To do this, the treasurer must sell his country's currency and buy the other country's currency. There is some risk of devaluation reducing the extra income, but this can be watched carefully or even hedged.

Raising interest rates results in a stronger currency because the demand for that currency rises. Sometimes costs of goods and services from the high-interest-rate country become prohibitively expensive on the world market. In 1993, interest rates in Germany were 3 to 4 percent higher than in the United States, and the value of the German mark became so high that German companies were losing much of their export market. Since then, the mark has weakened and then strengthened again as the German economy recovered.

I believe the value of goods and services sets an upper limit on how far exchange rates can get out of line. Once export markets start disappearing, the interest rates and the exchange rates should drop.

Governments can also try to influence rates. If a country feels its currency is overvalued, it can sell some of its currency for another currency. These moves generally are futile, as the amount of money in the market dwarfs the resources of a government to influence it.

## *Bank Charges*

The rates listed in newspapers are the "center rates" that apply to interbank transactions of $1 million or more. The banks create their profit by charging a buyer-seller spread that varies by the size of the transaction and the liquidity of the currencies being traded. They charge more for selling a currency than they will give when they buy it.

Banks make their profits by this buyer-seller spread. For large transactions, the spread is in the range of tenths of a percent. You can minimize this spread by either aggregating exchange transactions and making fewer, larger transactions or by negotiating a rate with your bank based on an annual volume. You should work only with major money-center banks, because small, local banks will buy from them and charge higher rates.

The following is an actual currency quotation for buying and selling Swiss francs:

- For $100,000: 1.2650–1.2750
- For $1,000,000: 1.2675–1.2725
- For $10,000,000: 1.2682–1.2718

This quotation gives the bank a profit of 0.39 percent (from the center rate of 1.2700) for a $100,000 transaction, 0.20 percent for a $1,000,000 transaction, and 0.14 percent for a $10,000,000 transaction.

## *Staying Current with Exchange-Rate Markets*

A global buyer with a large volume of purchases of foreign-built products must stay abreast of the exchange markets. This means more than just checking the exchange-rate figures in a typical daily newspaper. *The Wall Street Journal,* for example, has a daily column explaining the movement in the currency markets for the previous day and indicating the general outlook for the future.

The *Journal*'s predictions for the short term are not particularly good, but it does predict one- and two-year trends fairly well. Reading this column daily will help you understand the forces driving the currency markets.

## HOW MUCH CAN EXCHANGE RATES CHANGE?

While there have been days when major currencies changed 12 percent against the dollar, these big changes do not happen many days in a row.

**TABLE 11–2**
*Six-Month Change in Dollar Value, 1986–1995*

| Percent of times that dollar changed a given amount | Versus yen | Versus mark |
|---|---|---|
| –25 to –20% | 0.8 | 1.5 |
| –20 to –15% | 4.7 | 18.9 |
| –15 to –10% | 16.8 | 25.0 |
| –10 to –5% | 19.3 | 19.5 |
| –5 to 0% | 26.8 | 15.0 |
| 0 to +5% | 18.7 | 9.3 |
| +5 to +10% | 7.6 | 5.7 |
| +10 to +15% | 4.2 | 4.4 |
| +15 to +20% | 1.1 | 0.6 |
| +20 to +25% | 0.0 | 0.1 |

I developed some statistics of the change in the dollar's value against two major currencies (the yen and the deutsche mark) over the period from January 1986 through mid-1995. To do this, I compared the value of the dollar in those two currencies at the start of every week with its value 26 weeks later. This gave 471 data points for each currency. I picked six months because it is the practical limit of hedging, as will be explained in the next chapters.

From this data, I developed a distribution of changes, shown in Table 11–2. It shows, for example, that the dollar weakened against the yen between 10 and 15 percent, 16.8 percent of the time. The data show that the dollar has a reasonable probability of changing plus or minus 20 percent, but not a high probability of changing much more than that.

## Cost Changes Due to Exchange-Rate Changes

Exhibit 11–1 shows the effect of exchange-rate changes when you price your purchased product in a foreign currency. (This will be a recurring graph throughout the currency section.) Assume that you will be buying a product that costs 100 million yen, with payment due six months from the date you entered into the agreement. Assume also that the spot rate at the time you entered into the agreement was 85 yen per dollar. If the yen–dollar exchange rate does not change, you will pay $1,176,471 for the product (¥100M divided by 85). However, exchange rates can undergo

**EXHIBIT 11–1**
*Dollar Cost for 100 Million Yen (cost at spot)*

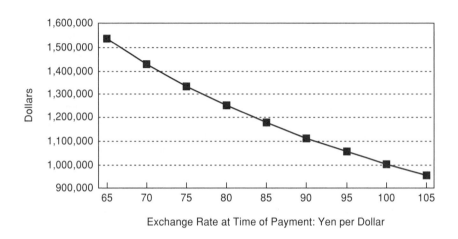

Exchange Rate at Time of Payment: Yen per Dollar

major changes. Exhibit 11–1 shows how many dollars you would have
to pay over a very possible range of spot rates on the date of payment.
Most companies find this level of uncertainty unacceptable and will take
steps to reduce the risk.

## RISK MANAGEMENT

There are several ways to reduce the risk of buying in a foreign currency.
You will need to use one or more of them when you are setting prices
in a foreign currency that is not pegged to the dollar. Three of these
methods are based on the financial markets. These are

- Forward contracts
- Futures contracts
- Currency options

There are also nonfinancial techniques, which are

- Risk sharing
- Escape clauses

Finally, there is the choice of buying in dollars, which also bears risks. The next few chapters will describe all these choices.

---

### Test of Understanding

1. You must pay an invoice of 300,000 Taiwanese dollars ($NT: New Taiwan dollars) on June 30. Ignoring bank charges, what will this cost in U.S. dollars?

2. Your bank quotes 25.758–25.898 Taiwan dollars for your transaction. How many U.S. dollars must you pay?

3. You are due a refund of 100,000 Mexican pesos for some returned goods. If the supplier refunds the money on June 30, what is the U.S. dollar equivalent, again ignoring bank charges?

4. You hear on the news that the German central bank (Bundesbank) has lowered the government-controlled interbank interest rate. What do you predict will happen to exchange rates? (Assume the market has not anticipated this move.)

5. The Bank of Japan is selling yen for dollars on the exchange markets. What are they trying to do?

Answers are in Appendix A at the end of the book.

---

## KEY POINTS

▶ Exchange rates are more sensitive to interest rates than to the value of goods and services. A higher interest rate increases demand for a currency and strengthens it.

▶ Trade balance has a longer-term effect on a currency. A country that exports more than it imports builds demand for its currency and thereby strengthens it.

▶ Banks make money buying and selling foreign currencies by charging more to sell a foreign currency than they give to buy it. This "spread" is tenths of a percent for transactions of more than $100,000.

▶ The risks of buying products in a foreign currency are too large for most companies, so hedging or other nonfinancial protective techniques are required.

*Chapter Twelve*

# Hedging for Fixed Dollar Costs

The uncertainty inherent in buying at the spot rate is unacceptable to most companies. The first inclination is to look for a way to make the cost certain. The futures and forward markets serve that function. They operate differently from each other but have the same effect. You can lock in an exchange rate today for payments that you have to make in the future.

## FORWARD CONTRACTS

Forward contracts ("forwards" for short) are financial instruments sold by major banks. They are listed in many newspapers and are available for all major currencies and some minor ones.

Please look at the newspaper currency listings in Table 11–1. Notice that the British pound, Japanese yen, and German mark all had forward rates quoted along with the spot rates. The spot rate for the Japanese yen was 84.710. The 30-day forward was 84.279, the 90-day forward was 83.637, and the 180-day forward was 82.615. The forward rates showed a weaker dollar or a stronger Japanese yen in the future.

Notice also that the forward rates for the British pound show a stronger dollar or weaker pound in the future. The spot rate that day was 0.6272, and the 180-day forward was 0.6306.

### Forwards' Difference from Spot

Forward rates are not anybody's prediction of the future. They are strictly a mathematical calculation based on interest-rate differentials between countries. As of June 30, 1995, Japan had lower interest rates than the United States. This caused its currency to sell at a premium (appear stronger) in the future, and the U.S. dollar to appear weaker in the future. The reason (somewhat simplified) is that a person who knows he or she has a debt in Japan in the future could take some dollars immediately, put them in a U.S. bank at a certain interest rate, take them out in six months, and then buy the yen. The buyer could also change money immediately, put it in a Japanese bank at a lower interest rate, and pay the debt when it comes due. However, because of the lower interest rate in Japan, the buyer would have to change more dollars now than he or she would have had to put in a U.S. bank.

---

### Forwards as a Predictor

While textbooks on currency explain that the forward rate is not a prediction of the future, I have a read a number of newspaper and magazine articles that claim that it is. There were enough of these articles that I decided to check out a test case.

I looked up the value of the spot rate and the 180-day forward rate of the German mark against the U.S. dollar at the start of every month for six years. I then calculated the percentage that the forward rate differed from the spot rate. For example, if the spot rate on January 1 was DM1.60 per dollar, and the 180-day forward was DM1.6160 per dollar, that calculates as a +1 percent difference between the forward rate and the spot rate.

I also calculated the actual change in the spot rates over the same six-month period. In the above example, if the spot rate was 1.60 on January 1, and 1.68 on July 1, that calculates to a +5 percent change in the spot rate.

Comparing the two figures quickly showed that the forward rate differential moved slowly and smoothly from month to month, and was never more than plus or minus 2 percent from spot. In contrast, the spot changes were as much as plus or minus 20 percent. This is clear evidence that the forward rate is not a predictor of the magnitude of change.

I also checked to see if the forward rate was a predictor of the direction of change. If the difference between spot and forward was consistently of the same sign (plus or minus) as the change in spot rates, the forward rate could be considered a predictor of the direction of change. In my test case, the directions of change were the same in 60 percent of the cases and different in 40 percent. This is not much better than being right 50 percent of the time, and I think it is a poor predictor.

The forward market takes this situation into account. It lets a trader keep his or her money in dollars until the date of payment, at a cost just under the cost of exchanging the money today and banking it today. This is called "interest-rate arbitrage."

The 180-day forward rate shown for yen is approximately 2.5 percent stronger than the spot rate. This indicates that the interest rate on six-month deposits in Japan was approximately 5 percent (per year) lower than the U.S. rates at the time of the quotation.

If the United States has lower interest rates than another country, the dollar will be stronger in the forward market than it is at spot. This is a purchasing advantage. If you price in foreign currency, you should get a lower base price. If you hedge a payment obligation to a supplier in a higher-interest-rate country by using a forward contract, you will get a better rate than the spot rate. When U.S. interest rates were lower than other industrial countries' rates (as they were in 1993), most major currencies sold at a discount on the forward market. This resulted in a "negative cost" for hedging through forwards.


## CASH FLOW

These rates are contract rates. If you need to pay yen 180 days from June 30, on June 30 you would sign a 180-day forward contract to buy yen. Your rate would be 82.615 yen for each dollar. You then know exactly how many dollars you would pay in the future.

No cash changes hands when you buy the contract. However, the contract is an obligation that can affect a company's credit. Your bank will want to be sure you have an adequate balance or credit line so that you will be able to pay the necessary amount in six months. The bank makes its money by charging a buyer-seller spread, as it does for spot rates.


## EFFECT OF FORWARDS

The graph of the dollar value of ¥100M in Exhibit 11–1 showed a change in dollar cost when exchange rates changed. Exhibit 12–1 shows the effect of a purchase of a forward contract. This contract is at the 180-day forward rate of ¥82.615 per dollar. The buyer knows how much the yen (and therefore the product) will cost in six months.

**EXHIBIT 12–1**
*Dollar Cost for 100 Million Yen (Cost at Spot and with Forward Contracts)*

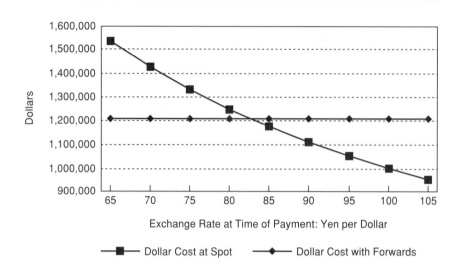

## Internal Forwards

Many companies find that they can create their own forward contracts internally, rather than deal with banks. If a company exports to another country and simultaneously buys products from that country, the sales department and the purchasing department can work well together.

The purchasing department:

- Wants to trade dollars for foreign currency to pay the supplier.
- Benefits from a stronger dollar.

In contrast, the sales department:

- Wants to trade foreign currency from customers for dollars to pay internal factories.
- Benefits from a weaker dollar.

The sales operations of a U.S. exporter benefit from a weaker dollar because they can either leave foreign currency prices fixed and get more dollars per sale, or they can reduce foreign currency prices and increase market share with no reduction in dollar revenue.

The complementary currency benefits allow companies to exchange currencies internally, rather than working through banks. With proper internal accounting, only the difference between flows needs to be exchanged externally. If a company needs to change $5M from dollars to yen and $4M from yen to dollars, only $1M needs to be exchanged at banks.

If your company currently changes more foreign currency to dollars than the other way around, you will reduce your company's total exposure by increasing your purchases in foreign currency. Your treasury department should welcome this.


## FUTURES CONTRACTS

Futures contracts have an affect similar to that of forward contracts, but they work differently. A buyer contracts to buy a number of standardized blocks of a foreign currency at a fixed rate on an established date. The difference is that futures are not bank instruments, but private contracts traded on an exchange. Contract sizes are fixed in blocks. Expiration dates are fixed on one specific date in the last month of every calendar quarter.

These are largely a vehicle for speculators, who trade them for profit and loss. The vast majority never result in an actual currency trade. However, they can be used for hedging, and can sometimes be cheaper than forwards for smaller companies.

You can find futures prices in major U.S. newspapers, such as *The Wall Street Journal, The New York Times,* and the *Los Angeles Times.* They are bought in brokerage accounts. Enough cash must be in the account to meet the margin requirements of the listing exchange, which is usually

**TABLE 12–1**
*Futures Prices for June 30, 1995 (Excerpts)*

**Deutsche Mark (CME) 125,000 marks; $ per mark**

|  |  |  |  |  |  | Lifetime |  |  |
|---|---|---|---|---|---|---|---|---|
|  | Open | High | Low | Settle | Change | High | Low | Open Interest |
| Sept | 0.7265 | 0.7275 | 0.7220 | 0.7254 | −0.0016 | 0.7450 | 0.6347 | 38,426 |
| Dec | 0.7264 | 0.7284 | 0.7250 | 0.7276 | −0.0016 | 0.7490 | 0.6580 | 1,905 |
| Mr 96 |  |  |  | 0.7294 | −0.0018 | 0.7505 | 0.6250 | 276 |

the Chicago Mercantile Exchange (CME). A brokerage commission is also required.

A listing of futures contract prices for the German mark is shown in Table 12–1. The table shows historical as well as current information. The key data is in the column labeled "Settle," which is the closing value for the day. This column shows dollars per mark, which is the opposite of the usage in this book. To convert to marks per dollar, divide the figure given into 1. For example, 1 divided by 0.7254 (the contract price for September contracts, approximately 90 days from the listing) gives 1.3785 marks per dollar. This is very close to the forward contract price of 1.3768 marks per dollar.

The futures contract prices will always be very close to the forward contract prices. Arbitrageurs will take advantage of any differences to move prices together.

Deutsche mark futures trade in blocks of 125,000 marks on the Chicago Mercantile Exchange. Other currency blocks available are:

- 12.5 million Japanese yen.
- 100,000 Canadian dollars.
- 62,500 British pounds.
- 125,000 Swiss francs.
- 100,000 Australian dollars.

At the time of writing, these blocks ranged in value from approximately $75,000 to $145,000.

---

### Test of Understanding

1. Were German interest rates higher or lower than U.S. rates on June 30, 1995?

2. What would a forward contract for 600,000 British pounds cost for delivery in 90 days, ignoring bank charges?

3. As a buyer, are you contracting to buy or sell the foreign currency?

4. The bank quotes you 0.6244–0.6324 for your transaction. How many dollars will your £600,000 cost?

Answers are in Appendix A at the end of the book.

## KEY POINTS

▶ Forwards and futures guarantee a fixed exchange rate.

▶ They do not predict the future.

▶ The forward exchange rate will be lower than spot if the other country's interest rates are lower than those in the United States. The forward rate will be higher if the country's interest rates are higher than those in the United States.

▶ If your company exports more to a country than it imports from that country, internal hedging with forward contracts will reduce the total currency exposure.

# Chapter Thirteen

# Currency Options

In Exhibit 13–1, below, you will see that buying at spot would have been the lower-cost choice if the dollar had strengthened. If the dollar had weakened, forwards would have been the lower-cost choice. The problem, of course, is that you cannot know what the dollar will be worth some months in the future. You don't know in advance which course will turn out to be better.

If you hedge with an instrument called a currency option, you can come close to the ideal solution. If the dollar weakens, you are protected against a price rise. If the dollar strengthens, you get a dollar cost decrease. However, like all good things, there is a cost.

**EXHIBIT 13–1**
*Dollar Cost for 100 Million Yen (Cost at Spot and with Forward Contracts)*

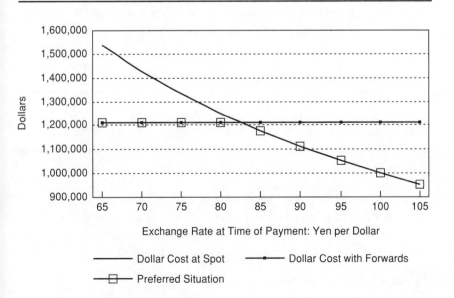

## OPTIONS DEFINED

An option is the right, without any obligation, to buy or sell a currency at a negotiated rate on a negotiated date. It differs from forwards and futures, which are contracts that both the buyer and seller must honor. A buyer takes an option to buy foreign currency (a "call" option). He or she exercises the option if the currency is cheaper at the option price than it is at spot on the day the option expires. This happens if the dollar has weakened below the option price. If the dollar has not weakened, the buyer just allows the option to expire.

The options used in purchasing are called "European style," which means they can be exercised only on the expiration date. American-style options can be exercised at any time by either the buyer or seller of the option. The bank that sold the option could close you out at any point, which would leave you without protection when you need it.

## OPTION MARKETS

Options are traded on the Chicago and Philadelphia exchanges, and are also available from major banks. The options available on the exchanges are for fixed dates and amounts, and I have never found one listed that was very useful. These are another vehicle for speculators.

A big money-center bank can quote you an option for most majoɪ currencies. You have flexibility for dates and amounts of currency.

## OPTION EXAMPLE

In this example, we are again entering into an obligation to pay ¥100M in six months, and today's spot rate is 85 yen per dollar. You buy an option for $60,000 that gives you the right to buy ¥100M for ¥85 per dollar (called the "strike price") on a date six months in the future. The $60,000 is called the "option premium."

Six months (less a few days) from now, you check the new spot rate and see if you want to exercise the option. If the spot rate is ¥80 per dollar and you can get ¥85 per dollar with your option, you would exercise the option. You can get more yen per dollar with the option than you can at spot.

However, if the spot rate is ¥90 per dollar, you would not exercise the option because you could buy more yen per dollar on the spot market. You exercise the option only when the dollar is weaker than your option strike price of ¥85 per dollar.

Exhibit 13–2 and Table 13–1 show the result. Your total cost is the cost of ¥100M plus the $60K option premium. If the dollar has weakened, you buy the ¥100M for $1,176,470 (¥100M divided by 85 yen per dollar). If the dollar has strengthened, you buy the ¥100M at the day's spot rate.

You are close to the desired situation of paying spot if the dollar strengthens and the forward price if the dollar weakens, but you have paid $60K for the privilege. This $60K is slightly more than 5 percent of the value of the option. This is typical of six-month option premiums during a period when currencies are volatile.

**EXHIBIT 13–2**
*Dollar Cost for 100 Million Yen (Option at 85 Yen per Dollar; $60,000 Premium)*

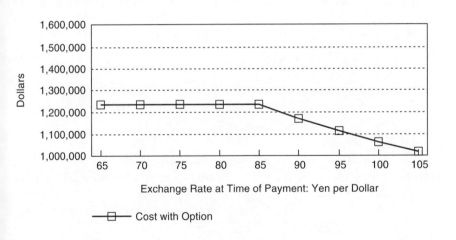

**TABLE 13–1**
*Dollar Cost for 100 Million Yen*

| Exchange Rate at Time of Payment | Cost at Spot (1) | Cost Using Option (2) | Exercise Option (3) | Total Cost (4) |
|---|---|---|---|---|
| 65.00 | $1,538,462 | $1,176,471 | Yes | $1,236,471 |
| 70.00 | 1,428,571 | 1,176,471 | Yes | 1,236,471 |
| 75.00 | 1,333,333 | 1,176,471 | Yes | 1,236,471 |
| 80.00 | 1,250,000 | 1,176,471 | Yes | 1,236,471 |
| 85.00 | 1,176,471 | 1,176,471 | Doesn't matter | 1,236,471 |
| 90.00 | 1,111,111 | 1,176,471 | No | 1,171,111 |
| 95.00 | 1,052,632 | 1,176,471 | No | 1,112,632 |
| 100.00 | 1,000,000 | 1,176,471 | No | 1,060,000 |
| 105.00 | 952,381 | 1,176,471 | No | 1,012,380 |

1. ¥100,000,000 divided by the exchange rate.
2. ¥100,000,000 divided by 85 yen per dollar, the strike price.
3. Exercise the option if the spot rate is less than the strike price.
4. $60,000 plus the lower of cost at spot or cost using option.

## OPTIONS COMPARED WITH SPOT AND FORWARDS

Exhibit 13–3 shows the total dollar cost of ¥100M when buying at spot, with an option at ¥85 and with a forward contract. In this hypothetical case the price of the forward is assumed to be ¥82.9 per dollar, about 2.5 percent below the spot rate.

This shows that options are never the cheapest choice. If the dollar weakens, you are better off with forwards. If the dollar strengthens, you are better off buying at spot. However, if your company believes that buying at spot is unacceptably risky, then options are the cheaper of the remaining two choices when the dollar strengthens.

## CONTROLLING OPTION COSTS

Options can be priced at almost any exchange rate. If the spot rate is ¥85 per dollar on the day you enter into the option, you could save money on the option premium by setting the strike price at ¥80 per dollar. At ¥80

**EXHIBIT 13–3**
*Dollar Cost for 100 Million Yen (Option, Forward, and Spot: Option at ¥85, Forward at ¥82.9)*

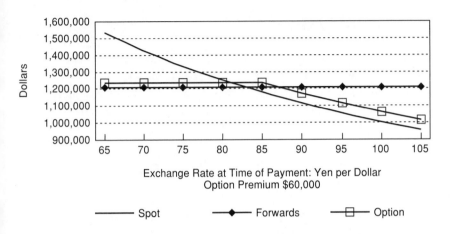

Exchange Rate at Time of Payment: Yen per Dollar
Option Premium $60,000

——— Spot          —◆— Forwards          —☐— Option

per dollar, the option is "out-of-the-money," because it is not worth anything until the dollar drops in value. (A strike price of ¥85 per dollar would be called "at-the-money" and ¥90 per dollar "in-the-money.") An option out-of-the-money costs less than an option at-the-money. It's a reasonable choice when you can tolerate some small increase in dollar cost but want insurance against major changes.

The total cost of an option depends on three things: the time into the future (expiration date), the strike price, and the volatility of the currency market. Unlike forwards, the price of which is set by interest-rate differentials, the price of options varies with market forces and will be most expensive just when you are most interested in buying them.

The ranges of costs of an at-the-money option are approximately 3 percent per quarter when currencies are volatile, and 1 percent per quarter when they are not.

## INTERNAL HEDGING

Unlike forwards, options cannot be used internally for hedging. The essence of an option is that the buyer of the option is not required to exercise it. If two departments of a company took opposite option positions, one would always refuse to exercise it, giving the other no protection.

## CASH FLOW

Unlike forwards, where there is no up-front cost, an option premium must be paid in advance. The premium can be significant, and unless there is a carefully thought-out strategy put forward to the finance department of your organization, approval will be difficult to get. There is more activity in hedging with forwards than with options.

---

*Test of Understanding*

You take an option to buy DM500,000 at a rate of DM1.7 per dollar. The option premium is $35,000.

1. What is the total cost of the DM500,000 on the expiration date if the spot rate that day is DM1.60 per dollar? Include the cost to buy the marks plus the option premium.

2. Cost if spot is DM1.70 per dollar?

3. Cost if spot is DM1.80 per dollar?

Answers are in Appendix A at the end of the book.

---

## *KEY POINTS*

▶ Options protect you against a weakening dollar, but allow you to take advantage of a strengthening dollar, at the cost of an up-front premium.

▶ The exchange rate guaranteed by the option is called the strike price. It can be almost any exchange rate.

▶ Costs are 1 to 3 percent per quarter to protect a spot exchange rate, and significantly less for exchange rates below the spot rate.

► Options cannot be done internally.

► A buyer takes an option to buy the foreign currency for dollars "European style."

► An option is exercised if the dollar has weakened below the strike price. The option is allowed to expire if the dollar is above the strike price at the time of expiration.

## Chapter Fourteen

# Hedging Practices

In the last few chapters, I implied that a company that is hedging would buy foreign currency via a forward, future, or option and use it to pay for parts. This is close to what actually happens, but not exactly right. Cash flows are too unpredictable to have hedges mature on the exact dates needed, and many small hedges are more expensive than one large one.

## BEHIND THE SCENES

What actually happens is that you pay your foreign currency bills by buying the foreign currency at the spot rate on the day you pay the supplier. You either buy it on the spot market or you buy it from your own company's treasury. In the latter case, accounting standards require you be charged spot rate. If the currency has changed in value from the time you issued the purchase order (or set a standard cost) you will have a price variance. If the dollar has weakened, you will have an unfavorable variance (parts were more expensive than anticipated) and if the dollar has strengthened, you will have a favorable variance.

The hedge activity takes place behind the scenes. There is a gain or loss on the hedge that offsets the purchased part variance. If you hedged with a forward contract, there will be a loss on the hedge if the dollar strengthens, and a gain if the dollar weakens. If you hedged with an option, there will be a gain if the dollar strengthens, but no loss if the dollar weakens because you do not exercise the option. Unless your treasury department needs the currency you buy in a hedge transaction elsewhere in the company, they will immediately sell it at the spot rate and convert it back to dollars.

### *Example*

Using the previous example, assume you order ¥100M in goods to be paid for in six months. You expect this will cost $1,176,471 based on today's exchange rate of ¥85 per dollar. You hedge with a 180-day forward contract for ¥100M. In this example, the forward contract rate is ¥82.9 per dollar. You expect your cost to be ¥100M/82.9, or $1,206,273.

Your production schedule changes, and you pull the order in by two weeks. You don't need to change the hedge. The invoice arrives and, for this example, assume the dollar has weakened to ¥75 per dollar. Your accounts payable department buys ¥100M at the spot rate, which costs $1,333,333 (100M divided by 75). You paid $127,060 more than you expected.

Two weeks later, the forward contract expires. You pay $1,206,273 and get ¥100M, according to the terms of the contract. If the exchange rate is still ¥75 per dollar, your ¥100M is worth $1,333,333. You convert it back to dollars, at a profit of $127,060. Your accounting system will credit this amount to the purchased part variance account, which now becomes zero.

The formula to calculate the gain or loss on a forward contract is

$$P_S = \frac{V_F}{S} - \frac{V_F}{C}$$

where $P_S$ is the gain or loss in dollars, $V_F$ is the amount of foreign currency you contract for, S is the spot rate on the day the hedge matures, and C is the original contract rate. $V_F/C$ is the amount of dollars you pay to close your forward contract. $V_F/S$ is the dollar value of the foreign currency you just bought. If $P_S$ is positive, there is a gain on the hedge. If $P_S$ is negative, there is a loss on the hedge.

If you hedge with an option, the above formula applies if you exercise the option. In this case, C is the strike price. Remember that there can be no loss on an option because you would not exercise it for a loss. The profit and loss formula does not include the option premium because you pay the premium at a different time. You pay it when you sign the contract.

Exhibit 14–1 shows how this works. In this example, you expected a product priced at ¥100M to cost $1,206,273 (the cost in yen divided by the forward contract rate of ¥82.9 per dollar). You hedge with a forward contract for ¥100M at ¥82.9 per dollar. When you pay the supplier, you pay with currency that you buy at the spot rate. This creates a favorable or unfavorable variance. You then have a gain or loss on the hedge that offsets the favorable or unfavorable variance of the part cost. The total cost is $1,206,273 regardless of the spot rate.

**EXHIBIT 14–1**
*Effect of Hedging with Forward Contract (Contract for ¥100 Million at ¥82.9 per Dollar)*

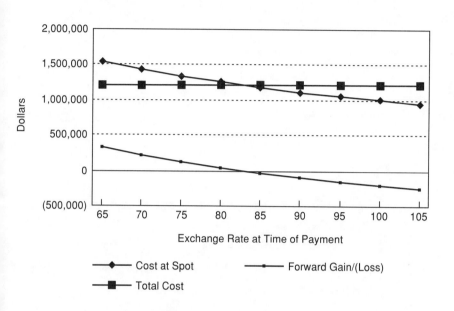

If you hedge with an option, a different graph results. You first have to determine if you would exercise the option. You exercise the option only if the spot rate is below the contract rate. You also have to account for the option premium.

Exhibit 14–2 illustrates this. The cost at spot is the same as in the previous graph. There is a fixed $60,000 option premium. There is a gain on the option if the spot rate is below the contract rate, which is ¥85 per dollar. The total cost is the total of:

• The cost of the product at the spot rate.

• Plus the option premium.

• Minus the gain on the option.

**EXHIBIT 14–2**
*Effect of Hedging with Options (Contract for ¥100 Million at ¥82.9 per Dollar)*

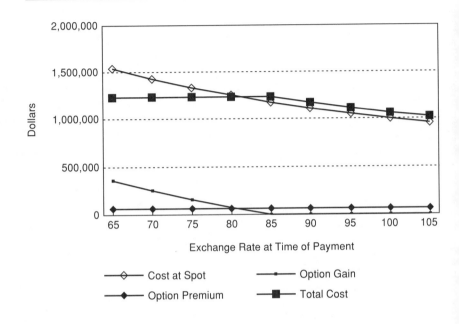

## TIMING OF HEDGING

Hedging for long periods of time is unrealistic. If you hedge with forward contracts, you might find yourself in a situation where you have to pay more than you would have paid at spot for an extended period. If, for example, you took a one-year forward contract at DM1.38 per dollar and the spot rate went quickly to DM1.60 per dollar, you would be paying more for the parts than you would have at spot. You do not want to be in this situation very long.

If you hedge with forward contracts, I suggest that you hedge three months' worth of orders. Every quarter, you would hedge the deliveries resulting from the next three months' orders. If your average lead time is two months and you pay 30 days later, this would result in a requirement to hedge cash flows for the period from three to six months away. If you hedge this way, you will know what items will cost in dollars before you place the orders.

If you hedge with options, you do not have the risk of paying more than you would have at spot. However, an option premium for a long period of time for a volatile currency is very expensive. You probably will not want to extend options any longer than six months.

Be sure that you have a plan in mind well before your hedges expire. All your prices will change, and you don't want to be surprised. If the dollar has weakened, you might want to find another supplier or renegotiate prices with the existing supplier. You also need an "escape" clause in your purchase agreement. This will be covered in a later chapter.

## FORECAST RISK

The biggest risk in hedging with either forward contracts or options is a result of uncertain forecasts of purchase volumes. However, the risk is not extreme and it is controllable.

If you hedge based on a forecast and your purchase volume does not match your forecasts, the gains or losses on the hedge will not exactly match the price variances on the parts. Unless your company either overbuys or underbuys chronically compared with forecasts, you have an equal chance of an unexpected gain or an unexpected loss. As a result you might wind up paying slightly more or less than you expected.

### Underbuying with Forward Contracts

You always have a gain on forward contracts if the dollar weakens and a loss if the dollar strengthens. (See the formula for determining the gain or loss from a forward contract on page 110.) If you buy less product than you forecast while you hedged with forward contracts, you will have bought a contract larger than you needed. Part of this gain or loss will not be offset by a price variance on the parts you buy. This uncovered gain or loss will remain in your price variance.

The uncovered gain or loss on a forward contract is approximately the percentage that you underbuy your forecast times the percentage that the dollar strengthens or weakens. For example, if you underbuy your forecast by 20 percent and the dollar strengthens by 15 percent, you have an uncovered loss of 3 percent of the value of the forward contract. (Twenty percent times 15 percent is 3 percent.) If the dollar had weakened by 15 percent, you would have had a gain of 3 percent.

The worst case is if you underbuy your forecast by 100 percent and actually buy nothing while the dollar strengthens. If you find yourself underbuying while the dollar is strengthening, you may decide to stop your losses by buying your way out of part of the contract. You do this by buying a second forward contract to sell the excess amount of foreign currency on the same date. This freezes your loss at its value on the date you buy the second contract.

## *Underbuying with Options*

You are going to exercise an option only if you can do so at a gain because the dollar has weakened. If you hedge with options and buy less product than you forecast, you will have an uncovered gain if the dollar weakens. You will finish with a favorable variance.

If the dollar strengthens, there will be no unexpected losses because you will not exercise the option.

## *Overbuying with Forward Contracts or Options*

If you buy more product than you forecast, you will have to pay spot rates for some of the product you buy and have no offsetting gain or loss on a hedge to compensate. If the dollar strengthens, you are ahead, but if it weakens, you will have to pay extra. Again, the amount of extra cost is approximately the percent that you underforecast times the percent that the dollar weakened. If the dollar strengthens, you have a similar unexpected gain.

Table 14–1 summarizes the gains and losses.

**TABLE 14–1**
*Summary of Gains and Losses*

| Effect of Missed Forecast | If Dollar Strengthens | If Dollar Weakens |
|---|---|---|
| If you buy less product than forecast | Forwards: Loss Options: No effect | Forwards: Gain Options: Gain |
| If you buy more product than forecast | Forwards: Gain Options: No effect | Forwards: Loss Options: Loss |

## *Exact Costs*

To determine the exact cost, you need to combine two items if you hedge with forward contracts and three items if you hedge with options. The first item is the dollar cost of paying the supplier with currency you buy at the spot rate.

The second item will vary with your hedge technique. If you hedged with a forward contract, you will have a profit or loss on that contract. Add a loss to your cost of paying the supplier, or subtract a gain from the cost of paying the supplier. If you hedged with an option, there cannot be a loss on the option because you would not exercise it. There might be a gain, however, and you would subtract the gain from the cost to pay the supplier.

The third item applies only to hedging with an option. You need to add the option premium to the combination of the other items.

## TIMING RISK

The other risk is a result of the hedge not expiring on the same day that invoices are due. The spot rate on those two days will differ, usually by a small amount. This means that the gain or loss on a hedge will not exactly equal the price variance on the purchased product. If the dates are no further apart than a few weeks, this risk is small. You can avoid this problem by arranging payment on one fixed date each month.

## BEST HEDGING PRACTICE

When you hedge a purchase you need to choose a course of action with two unknown variables. You do not know what the spot exchange rate will be when you actually pay the supplier, and most companies do not have very solid knowledge of the amount they are going to buy. It is impossible to pick the "one right hedge strategy" in these circumstances.

Some companies attempt to predict future currency values. Others do not. The management of these companies insists that one consistent strategy be followed.

## *Without Currency Prediction*

The decision tree in Exhibit 14–3 shows a process for choosing a currency strategy without attempting to predict currency values. It covers the complete topic, from pegged currencies through the choice between options and forward contracts.

## *With Currency Prediction*

If you make a currency prediction, the decision process will depend on whether your company allows you to buy unhedged with floating currencies. Most prudent companies will not permit this for significant purchases.

**EXHIBIT 14–3**
*Hedging Decision Tree*

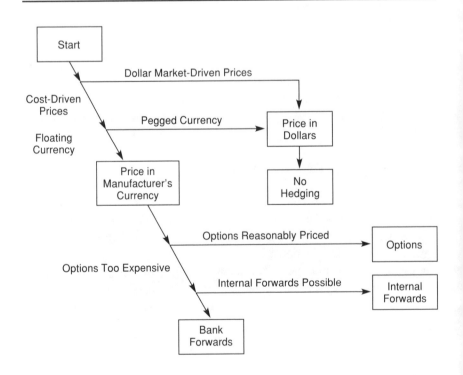

**Hedging required.**    If you must hedge, you have two choices: forward contracts or options. (Futures contracts may substitute for forward contracts for smaller companies.) If you expect the dollar to get weaker, forward contracts are the better choice. They are always cheaper than options.

If you expect the dollar to get stronger, the next question is "how much stronger?" If the expected strengthening of the dollar is more than the percentage cost of an option premium, then hedge with options. If the expected strengthening is less than the option premium, then forward contracts are the better choice.

**Hedging not required.**    In this situation, you have three choices: forward (or futures) contracts, options, and not hedging. If you expect the dollar to get weaker, then forwards are the best choice. If you expect the dollar to get stronger, then not hedging is the best choice.


## WHY NOT PAY DOLLARS?

You might be questioning the need for this complexity and uncertainty. Why not just price in dollars? There are several good reasons.

1. Your initial price will be higher. The supplier will have to take on the exchange risk and will charge a premium for it. If a supplier tells you the price will be the same in dollars or his manufacturing currency, there is money left on the table because the supplier must be paying for currency protection. The only case where this might not be true is if the supplier already imports more from the United States than it exports to the United States. In this case, sales to the United States priced in dollars would reduce the supplier's currency exposure.

2. Guarantees of fixed dollar prices are very suspect. A supplier is being unrealistic if it thinks it can hold prices constant in dollars in the face of a major dollar weakening. The supplier will get less and less of its own currency as the dollar weakens. It may have hedged against this possibility, but few companies will hedge more than three to six months. Pricing in dollars is no more certain than pricing in foreign currency with a hedge.

3. If the dollar strengthens, you are paying more than you had to. For a while, nobody might notice, but you will be aware and wonder if your competitors are paying less than you. Soon also, the supplier's competitors will be coming around and offering lower prices based on the stronger dollar. Finally, if the dollar gets really out of line, other channels of distribution can open. An agent in the supplier's country could buy the product for local currency and offer you the same part at the new spot rate, giving a lower dollar cost.

4. If you show unwillingness to price in the supplier's currency, the supplier will be more likely to put you in the "unsophisticated" category and want you to buy through his agents or representatives, rather than deal directly.

## MULTIPLE CURRENCIES

As international trade becomes more and more prevalent, fewer and fewer products are built using components from only one country. A Ford made in Canada or a Nissan made in Tennessee, for example, will have parts from many countries included in it. Lower-cost countries that are near major industrial countries (Mexico, China, and Poland, for example) are very likely to have assembly industries specializing in assembling components from their big industrial neighbors. There may be very little of your product's cost denominated in their currency.

European Union (EU) countries have a large amount of product from one another's countries in their assembled goods. The EU currencies also fluctuate against one another, despite their best efforts to keep the exchange rates constant.

Dealing with these products is conceptually simple but can get complex if you consider every aspect. During discussions with the supplier, you should determine the country of origin of the highest-cost purchased material. If these purchased materials come from countries other than the United States or the supplier's country, it may be necessary to have multiple currency pricing.

Part of the purchase agreement for the product will describe the major currencies involved. I suggest that you use no more than two currencies or you will become bogged down in complexity. For example, the agreement will state that "This product is assembled in Taiwan. Fifty percent of the manufacturing cost is in Japanese yen, and 20 percent is in U.S.

dollars." Because the Taiwan dollar is pegged to the U.S. dollar, only 50 percent of the product is at risk of fluctuating against the U.S. dollar. This means only 50 percent of the value of the product needs to be hedged, and the hedge currency is Japanese yen.

---

**Test of Understanding**

1. What is the gain or loss on a forward contract for ¥120,000,000 if the contract price is ¥85 per dollar and the spot rate at the time the contract expires is ¥95 per dollar?

2. You forecasted to buy 100,000 units of a product at ¥120 each, and you hedge with a forward contract. You actually buy 130,000, and the dollar increases in value by 15 percent from the time you contracted for the forward until you have to pay. Approximately how much more or less did your parts cost in total than you expected?

3. Same question as number 2, but you hedged with an option.

4. You expect to buy 100,000 units of a product at ¥120 each. You enter into a forward contract to buy ¥12,000,000 at ¥84 yen per dollar. Your schedule changes and you actually purchase 70,000 units. The exchange rate on the date you pay the supplier, which is the expiration date of the forward contract, is ¥95 per dollar. What is your total unit cost in dollars? (If you can do this one, you can discuss currency issues with your treasurer with confidence.)

Answers are in Appendix A at the end of the book.

---

## KEY POINTS

▶ You actually pay for the parts in foreign currency at the spot rate. A gain or loss is generated by a hedge behind the scenes and applied to a purchased-part variance account.

▶ The formula for determining the gain or loss from a forward contract is

$$P_\$ = \frac{V_F}{S} - \frac{V_F}{C}$$

where $P_s$ is the gain or loss in dollars, $V_F$ is the value of the hedge (expressed in foreign currency), C is the contract rate, and S is the spot rate at the time of expiration of the hedge.

► The formula for the gain from an option is the same formula. Options will not have losses.

► Forecast uncertainties create risk of uncovered losses on hedges. The risky area is if you overforecast purchases, you buy forwards, and the dollar strengthens. However, you have an equal chance of having an unexpected gain as a loss.

► The unexpected total gain or loss on a forward contract is approximately the percent of missed forecast times the percent the dollar changed.

*Chapter Fifteen*

# Nonfinancial Protection

The exchange rate protection that has been explained so far has all been via financial instruments. There are other, nonfinancial steps that companies can take to reduce risks.

## ESCAPE CLAUSES

Any purchasing agreement should have an escape clause for the party whose currency is not being used. If you are buying in foreign currency, you need an escape clause. If you are buying in U.S. dollars, the supplier will need an escape clause. If the supplier does not realize this, you might want to bring it up to avoid unpleasant surprises later. If the dollar weakens and you are pricing in dollars, the supplier will be able to get less of its own currency for your dollars and eventually will request relief.

Use your judgment about the relative power of buyer and seller in the situation and decide whether to bring up the issue. I usually raise the issue because raising it opens communication about the supplier's long-term plans. If the supplier is in a country whose currency is likely to appreciate, is he or she willing and able to take difficult steps to reduce costs? Can the supplier increase its foreign purchasing or move production to a less expensive location?

These escape clauses are useful only when there is an alternative source of supply or an alternative product available elsewhere. However, you should have such a clause in your contract in all cases. Under pressure, new sources or products might suddenly become more obvious. The clause can't hurt and might help. A good escape clause will set limits, define the data source, and describe what happens when the limits are exceeded. There is an example in Exhibit 15–1.

**EXHIBIT 15–1**
*Example of Buyer's Escape Clause*

---

This contract is based on 105 yen per dollar. If the U.S. dollar is below 95 yen per dollar, as reported in the New York edition of *The Wall Street Journal,* for 20 consecutive workdays, the buyer may request a price renegotiation. If the two parties cannot agree on an acceptable price, buyer may terminate the contract with reasonable payment for work in process, as defined in Exhibit X.

Note: This is a "high-context" (see Chapter 4) escape clause, suitable for countries such as Japan where buyers and sellers are expected to work out issues by mutual understanding. A low-context, more legalistic country would require a great deal more detail.

## RISK SHARING

Exchange risk sharing sounds simple. The buyer and seller agree to share the effect of exchange-rate changes. They usually agree to share the effect equally. If the dollar weakens 10 percent, the buyer would expect to pay 5 percent more dollars instead of 10 percent more. After watching several negotiations nearly come apart over details of this agreement, I realized it's more complicated than it seems.

The problem is that there is no mathematical formula that is totally fair to both parties. In the above example, if the buyer pays 5 percent more dollars, the supplier gets 5.5 percent less in his or her own currency. Suppliers are likely to object.

---

*Where Did That Half Percent Go?*

The built-in unfairness of risk-sharing clauses is not obvious. Here's an example. You have a contract to buy parts at $10.00 and the initial exchange rate is ¥100 per dollar. The supplier takes your $10.00 and converts it to ¥1000. If the dollar weakens 10 percent, to ¥90 per dollar, and your agreement was that you pay 5 percent more dollars, you pay $10.50. The supplier converts your $10.50 to ¥945. That's a decrease of 5.5 percent in the supplier's revenue.

## Formula, Local Currency Pricing

To avoid problems later in the contract, it's best to have a formula agreed upon between the parties. The formula that is fairest to both parties is the following:

$$\text{New Price in Manufacturer's Currency} = \frac{\text{Base Price} \times 2 \times R_N}{R_B + R_N}$$

Where

- Base Price is in the manufacturer's currency.
- $R_N$ is new exchange rate.
- $R_B$ is base exchange rate, the rate at the start of the contract.

In this formula, if the dollar weakens 10 percent, the buyer pays 5.26 percent more dollars and the supplier receives 5.26 percent less of its own currency. If the dollar strengthens 10 percent, the buyer pays 4.76 percent fewer dollars, and the supplier receives 4.76 percent more of its own currency. For the mathematicians among you, this is the geometric mean. This formula does not treat stronger dollars the same way it treats weaker dollars, but it affects both parties by the same percentage for any change.

If a supplier proposes another formula, the easiest way to evaluate it is to see the effect of a 10 percent weaker and stronger dollar. Check both what you pay and what the supplier receives in its own currency.

## Formula, Dollar Pricing

If you have a contract priced in dollars (as you might when you buy from a pegged-currency country) the equivalent formula is

$$\text{New Price in Dollars} = \frac{2 \times \text{Base in Dollars} \times R_B}{(R_B + R_N)}$$

## Risk-Sharing Details

You will also need to agree when price adjustments will occur. I believe quarterly is the optimum time. Changing every order or every month becomes administratively difficult. Also, I recommend that the adjustment take place on a time schedule regardless of how much or little the currency has changed. (The alternative is to ignore small changes.) I have seen too many cases where a buyer gets busy and forgets to check how much the currency has changed. Believe it or not, some suppliers will not tell you about a change that would reduce their revenues. It's best simply to make a note in your calendar just before the start of a quarter that it is time to adjust prices.

You also need to establish clearly what the rules are for determining the new exchange rate. You need to be clear on the source of exchange-rate data and the number of days of data that you will average to determine a new rate. The higher the number of days, the more stable the new rates will be. They will lag behind long-term trends, however. I think five days is a good compromise between stability and timeliness.

Finally, you need to establish whether prices change for open orders or not. A sample clause is in Exhibit 15–2.

## Risk Sharing, Pros and Cons

I don't like risk-sharing clauses because they seem artificial to me. If the dollar strengthens and the supplier's currency weakens, the supplier gets more of his or her own currency. There appears to be no reason for this.

**EXHIBIT 15–2**
*Risk-Sharing Clause*

---

Here is a sample risk-sharing clause.

"Prices will be adjusted at the start of each calendar quarter to reflect currency changes. The new price will be determined by the formula:

$$\text{New Price in Yen} = \frac{\text{Base Price} \times 2 \times R_N}{R_B + R_N}$$

where prices are in yen, the base price is as specified in this contract, the base exchange rate ($R_B$) is 85.60 yen per dollar, and the new exchange rate ($R_N$) is the average of the yen-per-dollar exchange rate as listed in the New York edition of *The Wall Street Journal* for the five-working-day period starting on the 20th of the month prior to the start of a calendar quarter. New prices become effective for all orders placed during the calendar quarter."

Also, the supplier will still have to build extra costs into his or her pricing formula to cover currency risk. Beware of suppliers who claim they are not doing this. They are either inexperienced or have a pad in their prices that you can uncover by further negotiation.

However, if your company is inexperienced in financial hedging and does not want to get started, risk sharing is an alternative that might bring the risks within an acceptable limit.

---

**Test of Understanding**

You have a contract for parts from the United Kingdom. The initial contract price is £1.20 per unit, and the base exchange rate is £0.65 per dollar. You have the above risk-sharing formula in your contract with the supplier. If the dollar strengthens 12 percent, what is the new price in pounds and what is your new dollar cost?

Answer is in Appendix A at the end of the book.

---

## KEY POINTS

▶ All contracts that may result in a price change due to currency fluctuations should have an escape clause.

▶ Risk-sharing clauses may be unfair to one side or another; evaluate supplier proposals carefully.

▶ Include both the formula and the clause defining its application in the written agreement between companies.

▶ The formula that treats parties equally is

$$\text{New Price in Manufacturer's Currency} = \frac{\text{Base Price} \times 2 \times R_N}{R_B + R_N}$$

where $R_N$ is the new exchange rate and $R_B$ is the initial or base exchange rate.

*Chapter Sixteen*

# Logistics

You do not need to be a logistics expert to be successful in global supply management. However, you do need to know enough to communicate with experts, to have a sense of what you can and cannot do, and to have a good idea of what various services cost. You are going to need this information even if you turn responsibility for shipping over to the supplier, because you will have to evaluate the supplier's shipping proposals.

## WHO SHOULD HANDLE THE LOGISTICS?

This depends on both the relative and absolute sizes of the buyer and seller. The larger shipper should be able to get the lower rates, but there is a lower limit after which rates do not drop much with size. If the smaller party is still a large shipper, the rates should be equivalent.

Most big companies who export a lot of products from the United States prefer to be responsible for the inbound shipment of purchased goods also. It increases the company's leverage with the shippers.

In addition, some major suppliers operate their shipping department as a profit center, and you can save money by handling the inbound logistics yourself. Be sure to check your supplier's proposed shipping costs against the rates you can get yourself.

## INBOUND FREIGHT FLOW

Exhibit 16–1 shows a typical inbound freight flow. It appears complex, but there are support organizations throughout the process.

Products will be picked up at a supplier's plant by truck or rail and delivered to the port (or airport) of exit.

**EXHIBIT 16–1**
*Typical International Product Flow*

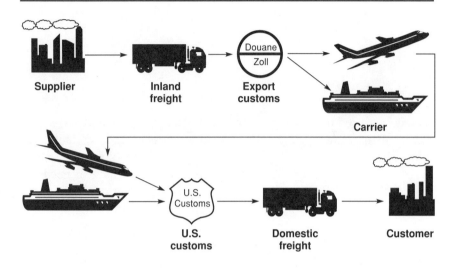

| Supplier | Inland freight | Export customs | | Carrier |

| U.S. customs | Domestic freight | Customer |

There the products must clear export customs. The export customs process is very perfunctory in most countries. Essentially, it is an export data collection effort plus a check to see if militarily restricted articles are properly licensed. However, there are a few countries where export customs is an issue. These tend to be countries with a history of large-scale income-tax evasion. Customs personnel in those countries will be concerned about overvalued exports, which is a way of moving profit out of the exporting country. Italy and Mexico are two examples.

After clearing export customs, the goods are put on a plane or ship (or trucked across land borders) and delivered to an area controlled by the U.S. Customs Service at a U.S. port of entry. Customs clears the goods and they are picked up by a local trucker or railroad and delivered to the buyer's location.

## SUPPORT ORGANIZATIONS

Fortunately, there are organizations along the way to coordinate a lot of these steps, at a cost that most people accept as reasonable.

## *Freight Forwarders*

Freight forwarders provide a variety of services. The main ones are to arrange the sea or air transport and to generate the shipping documentation. They also pick up goods at the supplier's factory and deliver the goods to the port. They send shipping details to the receiver of the freight, giving flight numbers or ship names and arrival dates. Once goods have cleared customs in the United States, they notify the buyer to pick them up. Some forwarders can also deliver the goods from the port of import to the buyer.

Freight forwarders get their revenue from freight carriers. They negotiate shipping rates based on their own total volume and sell space at a slightly higher price.

Some names of freight forwarders you may see around the United States are Kuehne & Nagel (German), Nippon Express and Kintetsu World Express (Japanese), and AEI (American).

## *Customs Brokers*

The customs broker is another valuable link in the chain. The broker might be integrated with the freight forwarder or might be an independent company. Brokers provide advice on customs classification, fill out the forms necessary to clear goods through customs, pay duties for the importer, and notify the importer that the goods are available. A later chapter of this book will cover customs brokers and customs issues in detail.

## AIR FREIGHT

Air freight is an increasingly popular option. It is best for light, small, and valuable goods. If these goods were to go by ocean, inventory costs would increase out of proportion to the freight savings.

### *Air Freight Rates*

Air freight rates are negotiable and confidential between the freight forwarder, buyer, and shipper. However, here is a range within which most large shippers will fall:

- Japan to U.S. West Coast: $4–$6 per kilogram (2.2 pounds).

- Other Asian locations to the U.S. West Coast: $2–$5 per kilogram.
- Europe to anywhere in the United States: $2–$4 per kilogram.

Rates are set by supply and demand, not distance. Air freight from the United States to Japan is approximately half the cost of freight from Japan to the United States because there is more freight eastbound than westbound.

Costs may also be based on physical volume, rather than weight. A light, bulky product takes up space on an airplane that could generate revenue. Carriers have devised a formula for "volumetric weight." The above prices are good for a kilogram of weight or 6,000 cubic centimeters of volume (365 cubic inches), whichever costs more. Thus, a one-kilogram product that is 12,000 cubic centimeters in volume would be charged as if it weighed two kilograms. Some carriers are starting to waive this requirement for large customers.

### Transit Times

In most cases, you should be able to pick up goods from a supplier in the morning and have them on a plane that day, or sometimes the next day. (Large countries with poor roads and limited airfields are an exception.) Once in a plane, the goods will be in a major U.S. airport the same calendar day (thanks to time changes and the International Date Line). Normally, they will clear customs the next day and be available for pickup. You should expect 0–2 days transit time from the supplier to your customs broker, and one day is typical. Exceptions are countries where there is a difficult transportation problem between the supplier and the airport, or no direct service to the United States.

### Reliability

Here there are differences between Asia and the rest of the world. For completely reliable transit times, your goods need to fly on a dedicated cargo plane. These planes can be operated by a passenger carrier (Japan Air Lines, for example), or by a freight line such as Federal Express or Nippon Cargo Airways. Passenger luggage takes priority over air freight on passenger planes. There is always some chance that your goods will be taken off such a plane to accommodate passenger luggage.

Air freight routes from Asia are much better developed than are routes from Europe or Latin America. If you are importing from Europe, you should first investigate air cargo flights before allowing your goods to go on passenger planes. Some have found it more reliable to truck goods across Europe to get on a dedicated cargo plane than to put goods on a passenger plane at an airport near the manufacturer.

## OCEAN FREIGHT

Most ocean freight travels in freight containers that also function as trailers for trucks. The container is loaded at the supplier and trucked to a seaport, where the container is lifted off its wheels and put into a container vessel. At the receiving port, it is put back on wheels and trucked to its destination. A container can also go directly onto a railcar. It will usually not be opened, other than for an occasional customs check of the goods inside.

### *Containers*

These containers come in two lengths: 20 feet and 40 feet. Width is 8 feet, and there are three heights: 8 feet, 8½ feet, and 9½ feet. Some countries, such as Japan, do not permit the 9½-foot-high containers because they will not fit under road bridges. (Japan was once accused of restraint of trade for having low bridges.)

The given dimensions are outside dimensions. For inside dimensions, take 8 inches off the length, 5 inches off the width, and 9½ inches off the height. In general, do not count on dimensions being consistent and do not count on being able to fill a container 100 percent. Containers do get dented in handling.

If your shipment is too small to have your own container and too heavy to go by air, you would ship by "less than container load" (LCL). A consolidator would put your goods with others' into a container, so that the goods still travel by container, not loose in a ship's hold. There are delays on both ends for loading and unloading from the container.

### *Carriers*

Ocean freight carriers break down into two groups: conference and nonconference. Conference carriers have organized themselves into a cartel and set standard rates that they pledge not to undercut. (Some negotiation is still possible, however.) These carriers tend to be better financed, easier to communicate with, more reliable in scheduling, and more expensive than nonconference carriers.

Nonconference carriers are not necessarily bad. Across the Atlantic, there is less difference between conference and nonconference carriers. If you are considering a nonconference carrier, you should check whether the ships adhere to a schedule, whether you can inquire in the United States as to the location of your freight and get quick answers, and how many stops the ship makes between pickup and delivery.

### Ships

Containerized vessels are very large. Some of the newer ones can carry nearly 3,000 forty-foot containers. It is surprising (to me) that about half of the containers will be carried above deck, where they are exposed to wind and water. The containers are enclosed, but not leakproof, so water can get in. You can specify below-deck storage, but it will cost slightly more, and you will never be sure you are really getting what you pay for. For this reason, you should be sure to pack your goods in packaging sufficient to ward off spray and drips from leaks.

### Ocean Freight Costs

Ocean freight costs are complex. There will be a list of prices covering export port charges, import port charges, currency adjustments, and fuel adjustments as well as the carriage itself. The total cost to move a 40-foot container from port to port and deliver to U.S. customs is approximately

- Japan to U.S. West Coast: $3,000–$4,000.
- Singapore to U.S. West Coast: $4,000–$5,000.
- Europe to U.S. East Coast: $2,500–$3,500.

The cost will vary (with conference carriers) with the product shipped. This cost applies regardless of how full the container is. A 20-foot container costs approximately 70 percent as much as a 40-foot container.

### Transit Times

Transit times are shorter than most people expect. Typical transit times are as follows:

- Japan to Oakland: 12 days.
- Singapore to Oakland: 20 days.
- Northern Europe to Boston: 15 days.
- Northern Europe to Oakland: 24 days.

A few new, high-speed, container-carrying vessels make the trip from Japan to California in eight days.

### Reliability

Again, there are differences between Asia and the rest of the world. Shipping services across the Pacific Ocean are well developed, frequent, and reliable. The ships are rarely late, and I regard a ship crossing the Pacific as more likely to arrive when advertised than the major parcel delivery services in the United States. If your goods are late going from the supplier to the ship, there is likely to be another ship the next day.

While the Pacific shipping lines are well developed, there are problems in the ports of some countries, in particular China and Thailand. These countries are industrializing faster than their infrastructure is developing. There are problems in obtaining containers, finding loading equipment, and even finding deep enough ports. You will need a really experienced freight forwarder in those countries.

Atlantic shipping has problems also. Ships are more likely to make several stops, either in Europe or in the United States, rather than moving point to point. This adds to delays and makes the service less reliable.

In addition, service is less frequent, so if you miss one ship, it may be a week before the next appropriate ship leaves. I have also found the communications from Europe about delays less reliable than they should be. Finding that goods missed a ship a week after it left is a frustrating experience.

### Optimum Shipment Size

Because you pay the same amount for a full container as a partial container, you will probably be inclined to ship full container loads and minimize freight costs. However, you should minimize the total of freight and inventory costs, and it might be cheaper to ship less than full loads of expensive goods.

There is a formula, much like the old EOQ (economic order quantity) formula, that gives the optimum shipment size:

$$\text{Optimum load in percent} = \frac{100}{L} \times \sqrt{\frac{2 \times C \times A}{i \times P}} \text{, where}$$

- $L$ is the number of units in a full container.
- $C$ is the cost of a shipment.
- $A$ is annual usage in units.
- $i$ is the inventory holding cost per year, expressed as a decimal fraction (e.g., 12% = 0.12).
- $P$ is the unit price of the product.

If this calculation gives a result of 100 percent or greater, ship full loads. If it gives numbers appreciably under 100 percent, consider a smaller container or air freight.

## INTERNATIONAL LOGISTICS AND JIT

Just-in-time (JIT) delivery can be achieved internationally with a reliable 2-day delivery cycle, as long as air freight is not too expensive and as long as dedicated air freighter service is available. It takes some fine-tuning and close cooperation among the freight forwarders, customs brokers, and carriers to achieve, but it can be done. Light, valuable items like electronic components are excellent candidates.

Some suppliers will propose that you achieve your JIT requirements by allowing them to stock goods in a warehouse near you. These goods would be used for emergency deliveries. Real JIT requires the supplier to manufacture just-in-time, not simply to deliver just-in-time. Using intermediate warehouses controlled by the suppliers is JIT purchasing, but it is not the same as buying from a JIT manufacturer. However, using a supplier-controlled warehouse in your country may still be a reasonable course of action.

## *KEY POINTS*

▶ Freight forwarders act as valuable intermediaries between shipping companies and buyers.

▶ Customs brokers act as an importer's representative in dealing with U.S. Customs.

▶ Logistics and customs are controllable processes. Delays in shipments or in clearing customs are symptoms of problems that can be fixed.

## *RESOURCES AND REFERENCES*

Hinkelman, Edward G., *Importer's Manual USA*. San Rafael, CA: World Trade Press, 1993. This book contains a good section on logistics and the documentation process for logistics.

# Chapter Seventeen

# Customs

Customs is another area where you don't have to be an expert, but you do need to know how to talk to the experts. You need to be able to ask questions in a way that will get them answered correctly. You also need to know about some opportunities for saving money and some common pitfalls to avoid.

## CUSTOMS BASICS

All goods entering the United States are subject to inspection by the U.S. Customs Service. They enforce customs laws and related laws having to do with import quotas for certain goods. While goods are subject to inspection, customs usually does not inspect them, provided the importer has a good record of following the laws and regulations. You need to maintain a good relationship with the Customs Service. If you don't, you will have many delays and much expense.

However, this doesn't mean that you should pay any more duties than are necessary. There are many totally legal ways to reduce or eliminate duties, and you should take advantage of all the possible duty exemptions and reductions. These exemptions will be explained after a discussion of some customs basics.

## DETERMINING DUTIES

You cannot get an accurate answer if you simply ask a customs official or broker "What is the duty on a power supply?" In order to determine a duty rate, you must consider four characteristics of an item:

- What is it?
- What do you use it in?

- Where was it made?
- Does it have U.S. content?

Duty rate determination starts with the first question. In 1991, the United States joined the rest of the industrial world by adopting the "Harmonized Tariff Schedule." All countries now classify products identically for at least the first six digits of an eight-digit classification system. They still may determine their own duty rates, however.

This schedule is called the "Harmonized Tariff Schedule of the United States." Table 17–1 is the reproduction of a tariff schedule section. This particular section is part of the classification for computers, and it illustrates several key points.

The eight-digit number in the left column is the classification code. It is referred to as the "HS Code," or sometimes the "HTS Code," meaning Harmonized System or Harmonized Tariff System. The next column is labeled "Statistical Suffix" and is a means by which the United States can add two more digits to an eight-digit code.

The third column contains the description. This is the official description of the part. In reading the descriptions, pay close attention to the level of indenting of the text. There can be as many as five levels of indenting. Indented text is a subset of the nearest text above that has one fewer levels of indenting. For example, classification 8471 has three subclasses on this page: Analog, Digital, and Other. You will occasionally see the term *NESOI,* which means "not elsewhere specified or included."

The next column is units of quantity, and the abbreviation "No." means "number."

The next three columns are "Rates of Duty." The first of those rates is the "General" rate, which is the rate that applies to imports made in "most favored nations." That sounds like an exclusive club but really includes almost all active trading partners. The countries excluded were usually Communist countries. There are fewer of these countries every day. In 1995 the excluded countries were Afghanistan, Cuba, Laos, Vietnam, Azerbaijan, Cambodia, and North Korea. Vietnam will probably gain MFN status in the next few years.

The next column is "Special." There are alphabetic codes for certain duty-exemption plans, which will be explained later. In this column, watch for asterisks next to the letter *A* or *E.* They are important.

The final column goes by various names. Its official name is "Original Statutory Rates," but it is also called "Column 2 Rates." These rates are

**TABLE 17–1**
*Harmonized Tariff Schedule of the United States (1995)*

| Head/ Subhead | Statis- ical Suffix | Article Description | Units of Quantity | Rates of Duty 1 General | Rates of Duty 1 Special | 2 |
|---|---|---|---|---|---|---|
| 8471 | | Automatic data processing machines and units thereof; magnetic or optical readers, machines for transcribing data onto data media in coded form and machines for processing such data, not elsewhere specified or included: | | | | |
| 8471.10.00 | 00 | Analog or hybrid automatic data processing machines | No. | 4.4% | Free (A,C, CA,E,IL, J,MX) | 40% |
| 8471.20.00 | | Digital automatic data processing machines, containing in the same housing at least a central processing unit and an input and output unit, whether or not combined | | 3.5% | Free (A,C, CA,E,IL,J, MX) | 35% |
| | | With cathode-ray tube | No. | | | |
| | 30 | Color | No. | | | |
| | 60 | Other | No. | | | |
| | 90 | Other | No. | | | |
| | | Other: | | | | |
| 8471.91 | | Digital processing units, whether or not entered with the rest of the system, which may contain in the same housing unit one or two of the following types of units: storage units, input units, output units: | | | | |

used for countries that do not qualify as "most favored nations." Notice that the General rates range from 0 to 3.7 percent, but the Statutory rates are 35 percent. This both explains why the current practice of reviewing China's MFN status annually is so troublesome to the Chinese government, and explains the importance to Vietnam of obtaining this status.

The general rule for looking up an item in the Harmonized Tariff Schedule is to look for it under its own name first. However, you should also look at the higher-level assemblies that the product will go into. You will often find the words "parts thereof" or "units thereof" in the description of goods. For example, classification 8471.99.60 is, in its entirety, "Units suitable for physical incorporation into automatic data processing machines or units thereof." This classification is duty free. You can find similar language in other classifications, such as those for aircraft, auto parts, machining equipment, and water pumps.

This seems to open up the possibility that anything that goes into a computer can come in duty-free. Unfortunately, the situation is not that simple. The rule is that if a part is described in two sections, the more specific description holds. A capacitor, for example, would be entered under a classification for capacitors. A sheet metal chassis designed to hold a disk drive can be classified as a part of a disk drive, though, because there is no classification for "chassis," except as a car part.

In another example, electric power supplies have their own classification code, 8404.04.80 (not shown here), with a 2.7 percent duty. However, they are also listed in section 8471.99.32, which is a power supply "suitable for physical incorporation into automatic data processing machines." This classification is duty-free. In this case, the latter description is more specific and will probably hold. When the results are ambiguous, you should consult a customs broker or attorney to see if you can interpret classification rules in your favor.

## SPECIAL TRADE PREFERENCES

The United States gives some countries and groups of countries duty-free treatment. These countries are indicated by code letter in the "Special" column. The duty-free programs are:

- A: Generalized System of Preferences (GSP)
- E: Caribbean Basin Initiative (CBI)
- IL: U.S.-Israel Free Trade Agreement

- C: Civil Aviation Pact
- J: Andean Pact
- K: Agreement on Trade in Pharmaceutical Products
- L: Uruguay Round Concessions on Intermediate Chemicals for Dyes

The letters "CA" and "MX" refer to Canada and Mexico. Their duty rates are set by NAFTA, the North American Free Trade Agreement.

## Generalized System of Preferences

The GSP program is designed to help developing countries' economies by allowing duty-free imports into the United States. (The European Union also operates a GSP program with slightly different rules.) Approximately 200 countries are eligible for this duty-free treatment. Russia and most of the emerging countries of eastern Europe are now eligible. The most important ones from a trade point of view are India and the members of the Association of South East Asian Nations (ASEAN), excluding Singapore, Brunei, and Vietnam.

The ASEAN countries are:

- Philippines
- Malaysia
- Indonesia
- Singapore
- Thailand
- Brunei
- Vietnam

In order to obtain duty-free treatment, the imported product must have at least 35 percent (of the sales price) come from the country in question. In the case of the ASEAN, the Caribbean Basin, and the Andean Pact countries, the content can be from any of the countries, again excepting Singapore, Brunei, and Vietnam. The first two countries—as well as Taiwan, South Korea, and Hong Kong—lost GSP eligibility a few years ago when the U.S. government determined that the countries had reached a stage of development where they no longer needed duty-free treatment.

The U.S. government will also exclude individual products whose imports to the United States from a normally GSP-eligible country become very large. You will often see an asterisk next to the letter "A" or "E" in the Special column. This asterisk means that one or more specific countries have been excluded from eligibility for that specific part. When you see an asterisk, you need to look up the specific HS code in another table that is part of the Harmonized Tariff system. For example, under 8471.92.32 (not shown), there is an asterisk next to the "A" in the Special column. Looking up that code in a table titled "General Note 4, subsection d" shows that these products from Malaysia and Thailand are ineligible for GSP treatment.

In order to obtain duty-free treatment under GSP, the manufacturer must fill out a "Form A," which certifies the amount of eligible content. This content can be local labor, overhead, and materials. Be sure not to permit a change in the process or the sources of components after you obtain certification, unless you certify again.

## North American Free Trade Agreement

The United States ratified NAFTA in late 1993. This treaty supplanted an existing treaty between the United States and Canada, the Canadian Free Trade Act.

Under NAFTA, duties and quotas (which will be discussed later) will be removed on all products of the United States, Canada, and Mexico that are traded among the three countries. Duties either were removed immediately, or are scheduled to be reduced on a 5-, 10-, or 15-year timetable.

Most of the duties between the United States and Canada had already been eliminated under the previous agreement, so the biggest impact is on duties for products traded between Mexico and the other two countries.

Duties for products from Mexico and Canada imported into the United States are described in an "Annex" to NAFTA, also published by the U.S. Customs Service. This document lists duties, describes the plan for duty elimination, and gives specific rules for determining whether a product is eligible for NAFTA treatment or not. It does this for each classification code.

Eligibility is automatic for any product built entirely of materials from the three North American countries. However, there are specific rules for every classification code that apply when there are goods from outside North America in the product. These rules prevent European and Asian manufacturers from opening "screwdriver factories" (that would simply assemble imported components) in Mexico. Generally, the rules require 45 to 55 percent North American content. Some products have more-detailed rules, which require certain components to come from North America.

This is a very detailed document, resulting from a complex treaty. Check very carefully before assuming that products from Mexico or Canada will be duty-free.

### Other Free-Trade Zones

"E" indicates the Caribbean Basin Initiative, under which the nations of the Caribbean receive duty-free treatment from the United States. "IL" is the U.S.-Israel Free Trade Agreement. Products from Israel are eligible for duty-free treatment by the United States if they do not enter into the commerce of a third country between Israel and the United States. (They are allowed to transit a free-trade zone of an airport.) "J" indicates the Andean Pact countries of Bolivia, Colombia, Ecuador, Peru, and Venezuela.

### Using the Harmonized Tariff Schedule

The Harmonized Tariff Schedule is a very long book, with a great deal of content. While it does have an index, it's often not very clear and written in difficult language. This particularly affects higher-technology parts. The Customs Service calls computers "automatic data processing machines," for example. Customs brokers must be very familiar with the schedule in order to obtain a license.

For a beginning user, I recommend buying the schedule in diskette form from the Government Printing Office. That will enable you, with the aid of a word-processing program, to look up words in the table easily, and will guide you to the proper section quickly.

## CLEARING CUSTOMS

Customs clearance procedures are usually simple, easy, and quick. While customs personnel definitely want to be sure you classify goods properly and pay duties, they are not keen on holding up goods needlessly. Unless you are randomly selected for an "intensive examination," or you are bringing in a product that is subject to careful scrutiny, customs clearance should take no more than one working day. Even in the case of intensive examination, if your goods and paperwork are correct, no more than two days should be needed.

## *Procedures*

If your goods arrive by air freight, some documents will be attached to the outside of the freight container. The plane crew will carry other documents. Your customs broker gets those documents from the airline, fills out one additional form, and presents the paperwork to the customs service at the airport. In the meantime, the goods are held in a bonded warehouse operated by the airline. Customs will release the goods to your broker (if they don't require an intensive inspection), and the goods are now available for pickup by your local trucker. Your broker guarantees payment of the duties later (although the ultimate legal obligation is yours).

Ocean freight procedures are somewhat different. No documents are on the outside of a freight container. Your freight forwarder (or the supplier's forwarder) must airmail or courier the documents to the port of entry so that they arrive before the goods. Port authorities will be annoyed if you do not have the documents ready when the goods arrive, and they can levy storage costs. In extreme cases, ports can auction your goods.

Procedures for clearance in ocean ports are similar to those in airports. In both cases, goods can be precleared. To do this, you need to work with a broker and freight forwarder who link electronically to the customs service through EDI (Electronic Data Interchange). This can save several hours.

## *Documents*

The required documents are:

- Evidence of the right to enter the goods into the United States. The ocean or air freight waybill serves this purpose.
- A commercial invoice that establishes the nature of the goods and their value.
- A packing list describing what is in each box of a multiple-box shipment.
- A customs entry manifest.

Your freight forwarder generates the freight waybills and the packing list. The supplier generates the invoice, unless you have an IPO (International Procurement Office) that bought the goods from the supplier and resold them to you. In this case the IPO generates the invoice. The customs broker generates the customs entry manifest.

### Costs

Both the U.S. Customs Service and the customs broker will charge you for entering goods into the United States.

**Customs Service charges.** The U.S. Customs Service charges a fee based on the value of the goods. It is currently 0.22 percent (except for goods from Canada or Mexico, the rate for which is being reduced to zero). There is a minimum charge of $25 and a maximum charge of $485 per entry. This is in addition to any duties that you must pay.

In addition, for goods arriving by ocean freight, the Customs Service will collect a "Harbor Maintenance" fee of 0.125 percent of the value of the goods. This is a boon to the port of Vancouver, Canada, and the truckers who deliver the goods across the Canadian border. The money they collect is earmarked for port improvements, such as dredging.

**Broker charges.** The broker will charge you a negotiated fee per entry. An "entry" is all the goods that arrive on one waybill (a logistics term meaning "a list of goods and shipping instructions"). In the case of air freight, this is called a "house air waybill" or HAWB. There is usually a higher charge for the first line item entered on the waybill and lower charges for successive line items. A typical fee structure might be $75 for the first item and $25 for each additional item.

This fee structure might lead you to put as many different items as possible onto one HAWB. This is a reasonable approach except that if one item on the HAWB is a problem in customs, all the goods on that HAWB are held until the problem is cleared up. To avoid this, use some judgment in choosing what items should be shipped together on the same HAWB. You can get multiple HAWBs on the same flight and it may be worth the extra broker expense to avoid problems.

While there is no requirement that you have a customs broker, it's the better approach when you have goods entering at multiple ports of entry. Otherwise you would have to have an employee at every port to represent you.

## KEY POINTS

▶ Duties vary based on the nature of the product being imported, the use of the product, the country of manufacture, and U.S. content in the product.

▶ Several trade pacts allow certain countries to export duty-free to the United States. The most important are NAFTA and the Generalized System of Preferences. Trade pacts usually have a requirement for some portion of the content of goods to come from the beneficiary country.

▶ In addition to duties, there are customs service charges and customs broker fees that must be paid when importing.

## RESOURCES AND REFERENCES

"Importing into the United States." U.S. Customs Publication 504. Contact: Superintendent of Documents, U.S. Government Printing Office, Washington, DC 20402. This 90-page booklet outlines the legalities of importing, as interpreted by the U.S. Customs Service.

*Chapter Eighteen*

# Customs and Logistics Practices

This chapter covers some "best practices" to minimize not only your costs of duties, but problems and delays in clearing customs. It also points out some problem areas that many inexperienced (and some experienced) importers encounter.

## REDUCING DUTIES

There are several ways to reduce duties or to get them back after you pay them. We have already covered Generalized System of Preferences, and you should check eligibility for GSP treatment whenever buying from a less-developed country. Israel, Canada, Mexico, and the Caribbean countries are also eligible for special treatment.

### U.S. Content of Assembled Products

If you purchase a product that has U.S. content, that portion of the product is not subject to duty. For example, if you import a product that has one-third of its purchase price made up of U.S. components, you can get the duty reduced by one-third.

This is not automatic. You must file an application, and the U.S. Customs Service will want details on the manufacturing process. Once you establish the process, you and the supplier must take care not to change it without amending the application.

There are other rules. One of the most troublesome rules is that the assembler must not modify the U.S. goods before putting them into the assembly. If your supplier imports a bracket from the United States for assembly into a sheet metal chassis, the supplier cannot drill holes into the bracket before assembling it into the chassis. It needs to be exported from the United States predrilled.

---

### Red Ink

We were importing a printer ribbon cartridge from Mexico. The cartridge assembly involved placing pre-inked ribbon from the United States in some U.S.-built plastic parts. The Mexican content consisted of the assembly labor and overhead and packaging. There was very little duty because of the high U.S. content. However, in trying to get the costs down, the supplier proposed inking the ribbon in Mexico.

Fortunately, someone was aware of the "no further work" rule and did some analysis before taking advantage of this offer. The ribbon would no longer be U.S. content because it would have been worked on in Mexico before assembly. The additional duties applied to the ribbon portion of the cartridge would have been higher than the savings resulting from transferring the inking to Mexico. Be careful!

---

## Duty Drawback

If you re-export something that you have imported, you can get 99 percent of the duty that was paid returned, as a "drawback." You have to know the amount of duty that was paid, show that you exported the imported part, and make an application. You can export the part as it was imported, or assembled into a higher-level product. The drawback process can take several months, but you will get the duty refunded.

In this case, the rules are reasonably user-friendly. You do not have to export the same physical parts you imported. You may export equivalent or identical parts. If for example, you have a domestic and a foreign source for the same part, you do not have to take care to put the foreign parts into the exported product. You can freely interchange the domestic and imported parts.

Determining the duty you paid is easy if you imported the parts yourself. However, if the part supplier imported the parts, paid the duty, and sold the goods to you in the United States, getting duty drawback becomes more complicated. It is possible for the supplier to certify to you the amount of duty it paid on the imports and to authorize you to get the duty back.

Some suppliers will do this, but others will not. By disclosing the amount of duty, they enable you to calculate the cost at which they bring the product into the United States. You can then calculate the supplier's markup for its U.S. distribution network. Many suppliers will not want this information made public.

This is a good reason to try to be the importer of record for as many imported parts as possible. It makes duty drawback easier.

### Other Countries' Duties

If you are buying an assembly offshore on a "turnkey" basis, with the assembler responsible for buying all the components in the assembly, be sure the assembler is not charging you duty or other taxes for products it must import.

Most countries offer a plan to either refund duties (U.S. style) or to allow temporary import without payment of duties. In the latter case, the supplier pays duties only on the parts it does not export.

## AVOIDING PROBLEMS

It is important to keep a clean record with the U.S. Customs Service. If you become recognized as a problem importer, your goods will be subject to extra scrutiny and there will be delays. Under a new regulation called the "Customs Modernization Act," Customs will be reducing the (already light) scrutiny it gives imported industrial goods. In turn, the Act places an increased burden on importers to show that they have used "reasonable care" in their importing decisions. Penalties for not showing reasonable care will be quite large. The obvious problem area is misclassifying goods or countries of origin. There are several other problem areas, however.

### Country-of-Origin Marking

Imported goods, or their packaging, must be marked with the country of origin. Generally, the product itself must be marked. If it is too small or would be damaged, the package can be marked instead. If you plan not to mark the parts themselves, you should discuss this issue with the Customs Service and obtain approval ahead of time, unless the reason is obvious.

The marking has to remain on the product or package until the goods reach what the Customs Service refers to as the "ultimate consumer." If you use the goods in manufacturing and assemble them into another product, you are the ultimate consumer. You can remove the marking at this point, so your finished product need not contain individual country-of-origin labels. However, if you sell the same parts as spares, directly to your customers, you need to keep the marking on or with the product.

Be especially cautious about hard-tooling country of origin into a part. It's not a good idea if there is any chance that the part will be visible in its final assembly. It might give the impression that the final assembly is from the country of the part and be seen as misleading by customs authorities. There is also a risk of the part being manufactured in another country at some point. Unless someone remembers to change the hard tooling, you will have mislabeled the part.

### Quotas

We actually live in a fairly protectionist country. There are limits (quotas) on the amount of certain products that can be imported into the United States. There are quotas on agricultural goods such as sugar, and on certain industrial goods. Fabrics and products made of fabric are the most common. Certain specialty steels have quotas also.

To import such products, you must have a "Visa." This has to be arranged in advance with the participation of the supplier. Unless you are an established importer of the product, they are difficult to get.

### Assists

Companies that buy goods that require tooling or design engineering find this issue particularly difficult. "Assists" exist when the cost of the part itself does not cover all the costs of manufacturing it. For example, most companies buy plastic mold tools separately from the plastic part itself. The cost of the part does not include amortization or depreciation of the tool, so the purchase price does not cover the true cost of manufacturing the part. Customs will want a duty payment based on the total manufacturing cost, so duty must be paid on the tool. This is true even if the tool never enters the United States.

Assists can also arise from design engineering charges such as the nonrecurring engineering (NRE) charges for application-specific integrated circuits (ASICs), unless the design was done in the United States. If you help the supplier in manufacturing the goods in any way, this is also an assist. However, quality testing is not regarded as manufacturing. This is a very gray area, and I advise you to seek advice from a broker or attorney if you are helping the supplier test products.

Duty on the assist is assessed at the same rate as the duty on the underlying part. You must declare the assist even if the duty rate is zero, so that the Customs Service can calculate U.S. import statistics correctly.

If the offshore tool is used to produce parts for the United States and for other countries, the value of the assist can be proportioned between each of the countries. If only 50 percent of the parts produced from the tool go into the United States, only 50 percent of the value of the assist needs to be declared to the U.S. Customs Service.

You need a consistent strategy for handling assists. The easiest way is to declare 100 percent of the assist on the first production part. There are other ways that are acceptable, generally involving amortizing the assist over less than a year.

---

*Innocents Abroad*

Here are a few ways that well-meaning people have run up against customs regulations.

- A visitor to South Korea bought athletic uniforms for his company volleyball team. He had them shipped back in a company shipment from South Korea to the United States. The company had no Visa for fabric goods, so the shipment was stopped. Unfortunately, there were production parts on the same HAWB, and some lines went down while customs tried to determine if there was intentional fraud going on. The uniforms had to be scrapped.

- Six souvenir flags caused almost as much trouble. They did not hold up production because they were on a separate HAWB, but it took six weeks to determine the nature of the fabric and the material of the flagpoles. Have fabric souvenirs sent to your home!

*(Continued)*

---

**Innocents Abroad** *(Concluded)*

- A circuit board had "Made in Hong Kong" included in the copper traces. New product managers transferred the board manufacture to Singapore and nobody noticed the Hong Kong label except the U.S. Customs Service.
- A product assembled in Europe included a cast-iron tube with "Made in USA" cast into it. The tube was quite visible and the customs service believed the importer was trying to mislabel the product as American.
- A buyer commissioned a $100,000 tool to build a plastic part that cost a dollar. The project was canceled, but the buyer decided to order one part from the toolmaker in order to check the quality of the work. The part was given a customs value of $100,001 (the tooling assist was amortized into the first part) and a 5 percent duty of $5,000.05 was assessed.

---

## Export Licenses

Despite the end of the cold war, there are still restrictions against exporting certain high-tech goods to some countries. Some products will require an export license in order to leave the manufacturer's country. Before an export license can be issued, an import certificate is needed from the importer. This simply states that the importer will not divert the goods illegally.

This process is actually quite simple. It is only a problem when it is a surprise, so be sure to check with the manufacturer of high-tech goods to see if he or she will need an export license.

## Product-Specific Requirements

Many products—specifically including food, clothing, electronics, household appliances, and drugs—have specific licensing and labeling requirements. Other government agencies develop these rules, but the Customs Service enforces them. Generally, the requirements are the same for domestic and imported products, so your company should be well aware of them. In most cases, there are exemptions for small-quantity imports for testing and evaluation.

## FOCUS YOUR IMPORTS

If you are going to be the importer of record, there are some steps you can take to make the process simpler and more controlled.

First, and most important, limit the number of ports of entry where you bring goods into the country. This will allow you to become better known to the Customs Service and to develop closer relationships with a few brokers.

Second, choose your customs broker carefully, with the intent of getting him or her involved in your business. Many duty-saving ideas can be developed by only someone familiar with the application of the products you import and the destination of the products you manufacture. Your customs broker should maintain a database of the appropriate classification for each of your imported parts.

## INCOTERMS

Incoterms are internationally recognized standard definitions that describe the responsibilities of a buyer and seller in a transaction. The International Chamber of Commerce developed these terms and has made them available in a convenient booklet.

There are 13 standard Incoterms, and there are three I recommend. (They are Ex works, Delivered Duty Unpaid, and Delivered Duty Paid.) Exhibit 18–1 shows where title transfers for each of the 13 terms. Each term must be followed by a geographic location such as a city or airport.

1. **EXW: Ex works.** (Example: Ex works Pusan) The buyer takes title when he or she picks up the goods at the supplier's factory, and is totally responsible for the shipment and duties.

2. **FCA: Free Carrier.** (Example: FCA Frankfurt Airport) The buyer takes possession at the airport or truck terminal at the port of export in the seller's country. Goods have cleared any export customs procedures.

3. **FAS: Free Alongside Ship.** (Example: FAS Yokohama) The buyer takes possession at the dock at the port of export. Goods have cleared export customs. Some Asian countries use the term "Ex-godown," but FAS is a better term.

4. **FOB: Free on Board.** (Example: FOB Yokohama) The buyer takes responsibility for the goods as they pass over the ship's rail during the loading process. This term is different from the accepted U.S. definition (and spelling) of F.O.B., which is widely used to mean "the place where title transfers." Try not to use F.O.B. in the usual U.S. sense in an international context. The correct terminology is "terms of sale."

5. **CFR: Cost and Freight.** (Example: CFR Oakland) The supplier arranges the freight and pays for it as far as the U.S. port of entry. However, title and the risk of loss belong to the buyer from the time goods go over a ship's rail in loading. You own the goods on a ship the supplier selected. I don't recommend this hybrid, confusing situation.

6. **CIF: Cost, Insurance, and Freight.** (Example: CIF Galveston) The supplier arranges the freight and buys insurance for the goods as part of the sales price. Title and risk transfer once the goods clear a ship's rail while being loaded. This is another confusing situation, with the added disadvantage that the seller may not be purchasing the most reliable or cost-effective insurance.

7. **CPT: Carriage Paid to.** (Example: CPT O'Hare Airport) This is similar to CFR but is used for air, truck, or "intermodal" (containerized) transport. Title transfers when goods are loaded into a container, and the risk of loss is the buyer's. However, the seller chooses and pays for the carrier. Again, I don't recommend this one.

8. **CIP: Carriage and Insurance Paid to.** (Example: CIP Laredo, Texas) This is similar to CIF, but is used for air or truck transport only.

9. **DAF: Delivered at Frontier.** (Example: DAF San Ysidro) Goods are delivered to a border by truck or rail. They are cleared for export but have not cleared import customs. Title transfers at the border.

10. **DES: Delivered Ex Ship.** (Example: DES Miami) The seller pays freight costs to a port of import, and title transfers in the ship on arrival. The buyer is responsible for unloading the freight and clearing customs.

11. **DEQ: Delivered Ex Quay.** (Example: DEQ Galveston) The seller is responsible for having the ship unloaded at the port of

entry. This term must be appended with "Duty Paid" (not recommended) or "Duty Unpaid." Title transfers after the ship is unloaded, either before customs (Duty Unpaid) or after customs (Duty Paid).

12. **DDU: Delivered Duty Unpaid.** (Example: DDU Colorado Springs) The seller delivers the goods to a location specified by the buyer, which is different from the port of import. Title transfers at this location. This involves transport of goods that have not cleared U.S. customs. You can arrange this with proper customs supervision and bonding, but you must clearly establish whether the buyer or seller will be handling the administrative aspects.

13. **DDP: Delivered Duty Paid.** (Example: DDP Buyer's Factory, Columbus, Ohio) The seller delivers the goods to the buyer with all duties paid. Title transfers at your location.

**EXHIBIT 18–1**
*Typical International Product Flow*

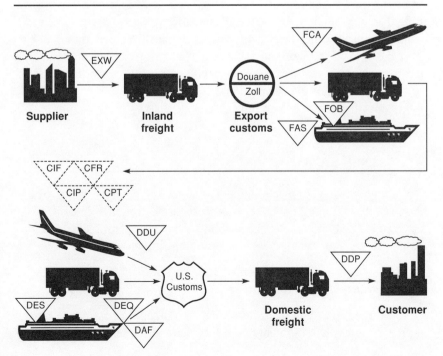

## *Selection of an Incoterm*

The appropriate term will depend on whether your company is handling its own freight and customs work. I recommend not using any term that allows the seller to buy insurance for you. I also recommend that you handle the import details and duty payment yourself. This makes the best terms Ex works if you handle the freight and customs, Delivered Duty Unpaid if you handle customs only, and Delivered Duty Paid if you handle neither customs nor transport.

It is a good practice always to get an Ex works quote in the currency of the manufacturer so that you can examine the total cost picture of the product, the logistics, and the duties separately. This shows active attention to costs.

## *Incoterms and Duties*

Your selection of an Incoterm affects duty costs because sometimes some of your freight costs are dutiable. International freight is not dutiable, but the cost of domestic freight in the seller's country may become part of the total cost of the part when calculating duties. If the buyer pays the foreign inland freight directly, as would occur with Ex works buying, that freight is not dutiable. However, if the seller pays that freight, as would happen with all other terms, the foreign inland freight is dutiable.

If the cost of the international leg of the transportation is included in the cost of the goods (as would occur with any Incoterm starting with "C" or "D"), the cost of the international freight must be subtracted from the total cost of the part before duties are calculated.

## INSURANCE

You should make your own judgment as to your company's ability to cover the loss of a shipping container or a missing air freight shipment. If you use established, reputable carriers and forwarders, problems are rare. However, most prudent buyers will have some kind of insurance coverage.

## *Blanket Coverage*

If you do feel that you need insurance coverage, it's best not to pay for it on a shipment-by-shipment basis. If you are a frequent shipper, you can get a blanket policy that will be priced according to the amount of freight you ship. The premium will also vary based on the deductible amount, which you can set at the maximum loss you can reasonably sustain.

### General Average

Maritime law has a long history, and some practices survive that seem to make no sense today. One of those practices is the concept of "General Average." A ship's captain, if he believes his ship is in danger and lightening the load will help save it, may order cargo jettisoned. Under General Average, all shippers who have goods on the ship must share in the cost of the goods thrown overboard. In addition, if a ship is lost or damaged, shippers can be held responsible for paying for the replacement or repair of the ship.

While I find it hard to believe that a reputable shipping company that is interested in a long-term relationship would invoke General Average, they have the right to do so. If you buy insurance, be sure it covers General Average.

## KEY POINTS

▶ Duties can be reduced according to the proportion of U.S. content in a product.

▶ Ninety-nine percent of duties can be refunded if an imported product is re-exported. It can be re-exported in the same condition it was imported or assembled into a higher level product. If the buying company was the importer, drawback is simple. If another company was the importer, that company can pass drawback rights to the buying company.

▶ A country of origin label must be affixed to important products or to their packaging. This label must stay with the goods until they are assembled into another product or are delivered as-is to another buyer.

▶ "Assists" occur when the price paid for a product does not cover the full costs of making it. Molded plastic parts for which the tool is paid separately are an example.

▶ Use a standard "Incoterm" to define the terms of purchase of a product. Recommended terms are Ex works, Delivered Duty Paid, and Delivered Duty Unpaid.

## *RESOURCES AND REFERENCES*

*Incoterms 1990.* New York: ICC Publishing Corporation. A 200-page booklet in English and French that describes each of the Incoterms in great, and necessary, detail.

## Chapter Nineteen

# Paying the Supplier

Paying a foreign supplier can be extremely simple or it can be very complex, risky, and expensive for the buyer. Payment methods range from modern electronic transfers to methods developed in the banking industry in the late nineteenth century that still survive today.

## LETTERS OF CREDIT

Some suppliers, unless they know you well, will not want to give credit terms. They will try to get you to use "letters of credit." These are instruments created by the banking industry that serve the banks and the sellers. They are something to avoid if you can.

Essentially, a letter of credit (LC) guarantees payment to a supplier if the supplier presents appropriate export documents to a bank in its country. The buyer specifies what is to be exported and to whom, the price, and the last shipment date. If the seller presents documents that show that all the conditions have been met, the bank in the supplier's country pays the supplier. Your bank in the United States will have transferred your money to the foreign bank before payment.

There are more complicated versions, such as those that result in paying the supplier a defined number of days after shipment, rather than on the date of shipment. You can add additional requirements to the documents required for paying the supplier. You can require a certificate of inspection by your representative, for example.

Banks developed this method of payment when travel was slow, expensive, and difficult. Mail and payments moved with the speed of a ship or a train. Buyer and seller might never meet each other, and there was no reliable credit information about either party. If you use the modern purchasing techniques explained in this book, you should not need to use such antiquated payment techniques.

## *Complexity*

The process of generation and payment of letters of credit is a bureau-cratic maze. The chart in Exhibit 19–1 shows the process of combining letters of credit with a payment "draft." (A draft is a demand for payment by a seller.) All the activity shown in Exhibit 19–1 consists of people looking at pieces of paper. They will be checking for compliance with the terms of the letter of credit and also for any discrepancies between the documents. The more documents there are, the higher the chances of a discrepancy.

This raises another concern I have with the LC system. The banks set themselves up as a neutral party between buyer and seller. As such, they are in a quasi-judicial role and seem to have no clear customer. They pride themselves on attention to fine detail. Often, however, they do this more as a protection for their own interests than in the interest of a customer. This makes them difficult to deal with.

**EXHIBIT 19–1**
*Letter of Credit and Draft Document Flow*

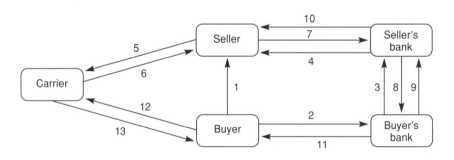

1. Place order
2. Arrange LC and payment
3. Send copy of LC
4. Send copy of LC
5. Ship goods
6. Send copy of bill of lading (BL)
7. Send draft and BL

8. Send draft and BL
9. Return payment and draft
10. Pay seller
11. Send copy of BL
12. Give carrier copy of BL
13. Give goods to buyer

## *Cost*

Letters of credit cost between 0.5 and 3 percent of the amount of payment. In addition there are charges in the $50 to $100 range for any changes to the letter's terms. If you change a quantity or a shipment date, the bank collects a fee. Historically, buyers pay these expenses. Every time you or the supplier needs to make a change, the bank must modify the letter of credit, and the bank will charge you.

In addition, the money to cover a letter of credit must be in your bank or in your line of credit. This ties up your capital from the time you open the letter.

A good rule for a buyer is never to pay an intermediary more than it would cost to do the job yourself. (More practically, never pay more than the opportunity cost of doing it yourself.) On a $10M transaction, an LC might cost $75,000 to $300,000 dollars. If your supplier is insisting on an LC, consider that it would cost you only a fraction of the LC cost to fly to the supplier's country and pay the supplier personally. This would also lead to a closer relationship with the supplier. Use this rule when dealing with banks on LC costs.

## *Supplier Bias*

Letters of credit offer complete protection to the supplier. From the time the supplier receives the LC, he or she is totally guaranteed payment. The supplier does not have to accept cancellations, quantity changes, or schedule changes. As long as the supplier presents documentation to the bank that shows it met the terms of the LC, it will receive payment. This is my biggest objection to the LC. Of course you should pay bills when due, but the inflexibility of the system does not match modern industrial needs for frequent schedule and quantity changes.

LCs offer minimum protection for the buyer and maximum protection for the seller. This is why sellers will often try to insist on a letter of credit. They might even claim it to be a matter of law. However, I know of no country that requires letters of credit. You should check out any claim that an LC is required by law with a local law firm. South Korea does not permit credit terms for exports, but even this country will allow other methods of guaranteed payment.

## DOCUMENTS AGAINST PAYMENT

This is another documentary payment method developed by the banking industry. "Documents" are the papers that prove you own the goods and allow you to import them. Your designated bank in the seller's country receives the documents and pays the supplier. It then sends the documents to you. In effect, this is a COD transaction.

This instrument costs less than a letter of credit, generally in the range of 0.5 to 1 percent. This payment method is slightly less biased in favor of the seller, as the buyer can refuse to accept the goods by refusing to pay. This gives you some ability to cancel or reschedule orders. Of course, if your relationship with the supplier requires you to use this method of making a schedule change, your relationship is in trouble anyway.

South Korea will accept this method of payment, which is a slight improvement over LCs.

## WIRE TRANSFER

The easiest and most convenient method of paying the supplier is through a wire transfer. You need to know the supplier's bank and account number. You simply instruct your bank (or your company treasury department) to make a wire transfer for a certain amount of a particular currency to the seller's bank.

You should work with a major bank that is a member of the SWIFT network, when making these payments. The SWIFT network is a group of banks that use a standard, efficient, and safe method of wire transfer. A bank will charge you around $35 for the transfer. They will also make a small profit on any foreign-exchange transaction if you are paying in a currency other than dollars. This profit will be in the tenths of a percent, and you should try to negotiate this spread with your bank based on annual transaction volume.

## OBTAINING CREDIT TERMS

This is your best solution. If you work for a reputable, ethical company that pays its bills on time, there is no reason that you should not receive credit terms from the seller. It will be easier for you to receive these credit terms if you have some legal presence in the supplier's country, such as a sales office or an IPO. You might be able to use the credit umbrella of your representative in the seller's country to obtain these credit terms more easily.

At the very least, a supplier whom you have met, whose products you have qualified, and who believes that you will be receiving monthly or weekly shipments, should give you one shipment's worth of credit. You would pay for the first shipment before the supplier ships the next. This is possible even with ocean freight, because documents arrive before the goods.

Your ability to negotiate credit terms successfully depends on your relative negotiating power versus the supplier's. You should let any supplier who is refusing to give you credit terms know that you are unlikely to do business on an LC basis. However, if the supplier offers an extreme competitive advantage that you can't obtain elsewhere, you might have to use LCs occasionally.

In this case, the supplier should agree to grant credit terms at some defined point in the future. You could agree, for example, that at the start 75 percent of the payment would be by letter of credit and 25 percent by wire transfer. If you pay on time, the percentage of payment by wire transfer would increase over time to 100 percent.

If you are with a big company that is negotiating with a smaller company, it may be worthwhile to offer to pay some expenses of a credit check and a consultation with a U.S. law firm in the supplier's country to reassure the supplier of your creditworthiness. This is better than paying with letters of credit.

### Payment Terms

Payment terms vary widely from country to country. Don't automatically offer "net 30 days" without knowing what is typical in that country. You may have a negotiable point in your selection of terms.

The typical practice in southern Europe is to pay closer to 90 days. In Japan it is typically 60 days, and in northern Europe it is generally 30 days or shorter. The American Chamber of Commerce in the supplier's country can give you typical payment terms. If possible, get credit terms that are longer than the shipping transit time. This way, you will have received the goods before you pay the supplier.

## KEY POINTS

▶ Buyers should avoid using letters of credit. They are biased in favor of the supplier—expensive and bureaucratic. Reasonable suppliers with whom a creditworthy buyer has taken the time to develop a relationship will not require them.

▶ Try to get payment terms that are normal in the supplier's country, and pay by a wire transfer.

## RESOURCES AND REFERENCES

"Guide to Documentary Credit Operations." New York: ICC Publications. An introduction to letters of credit and other documentary collection methods.

# International Procurement Offices

I have mentioned international procurement offices (IPOs) in various sections of this book. They are an excellent solution to the difficulties of distance, culture, currency, and logistics. Many medium- and large-sized companies set up one or more of these offices in regions where their suppliers concentrate. The largest companies set up networks of these offices around the world.

## SERVICES

IPOs provide a wide variety of services. They generally attempt to be a remote extension of the buyer's purchasing department.

### *Research and Quotes*

One of the most important functions of an IPO is to identify good suppliers and locate sources of needed products. This gives buyers at home the capability of "shopping the world" when they are seeking a new source or looking for opportunities for cost reduction.

To do this effectively requires engineering talent. The IPO must have sufficient technical capability both to recognize good suppliers and to convince the buyers and engineers at the purchasing location that the supplier will meet their needs. These engineers need experience in manufacturing the products that the company is likely to buy. More than 50 percent of the staff of most successful IPOs have engineering degrees.

### *Negotiation Support*

Having someone on your side who is familiar with the language and culture of the seller can be extremely valuable in negotiations. The IPO staff can help explain puzzling situations to you. In addition, remember that you may occasionally puzzle the supplier as much as the supplier puzzles you. The IPO is on your team and will be able to explain your behavior in terms that the supplier will understand and in a way that advances your negotiation position.

It is also a good idea to have someone on your team who speaks the supplier's language. If you don't, the supplier can easily have side discussions and coaching going on with you in the room. The other side's negotiating capability is multiplied because their people can speak openly in the meeting without your knowing what they are saying. Third-party translators could also help here, but they are generally expensive and will not understand your company's needs as well as an employee would.

Another form of support is to provide the legal umbrella that makes any contractual and credit arrangements purely domestic in the supplier's country. This eliminates the difficulties of international contracting and permits discussions to focus on true business issues.

Finally, IPOs can offer the simple but vital help of travel support. Just getting from a train station to a supplier's plant or from an airport into town can be difficult in some countries.

### Supplier Management

Supplier management primarily encompasses maintaining quality, delivery, and price competitiveness. This is one of the most important roles of the IPO. The IPO is geographically close to the supplier. Someone from the IPO can usually be at the supplier's facility in a few hours. This helps overcome the normal reaction of a buyer in a problem situation to wait for "one more shipment" before taking a long, expensive plane flight to see what is going on.

You should expect the engineers in an IPO to put together a quality assurance program for each product and supplier that the IPO manages. This program can range from statistical process control down to emergency source inspection. If the program is effective, the chances that you will be the person discovering a faulty shipment are minimized.

The IPO is also in a position to measure the supplier's shipping performance effectively, because the staff can easily find out when goods are shipped. This information is difficult to discover across an ocean because the performance of the shipping network becomes confused with the supplier's performance. A good supplier may appear poor due to the shipper's performance.

The IPO can also influence the supplier's delivery performance by sharing forecasts and other valuable information. A supplier should not be isolated from regular communication on how the business is developing, and suppliers of any critical component should be receiving forecasts. These could be sent by fax or electronic data interchange (EDI), but personal contact is more effective.

---

### Global Price Management

Earlier we discussed worldwide, dollar-based commodities. These commodities generally have the same price around the world. The question is how to keep them the same by constantly adjusting the higher prices downward. An international procurement office network is an ideal way. The same part can be bought at market price in Asia, Europe, and North America.

Your IPOs can quickly adjust the difference by pointing out to the salesperson in the higher-priced region that his or her region will be losing business if the price is not brought down. This technique can be applied to multinationals as well as to local companies. Asian companies operating in Europe should be expected to meet the same prices there as in Asia. (Duties and freight may be an issue at first, but if a manufacturer starts producing in Europe, the European market price of a commodity product should match the worldwide price.)

The human factor makes this work in most companies. The German sales manager of a Japanese company will want to have the sale made in Germany and will often fight openly or behind the scenes to have parts available to key customers at the price they want to pay. Few companies are able to control this tendency.

---

Finally, the IPO is in a position to understand the economic situation in a country and to know when to push for price reductions. Since an IPO's existence depends on having a competitive supplier base, an IPO is extremely motivated to pursue cost decreases.

## Order Management

Some IPOs issue purchase orders to suppliers, and others merely act as agents for orders from the United States. I have found that it is much easier for an IPO to be effective in supplier management if it issues the purchase orders to the supplier. This increases the perceived and the actual power of the IPO, because the IPO has the legal and cultural power of the buyer.

Quality programs involving source inspection are particularly difficult to operate without the power of a purchase order. Surprise shipments at unusual hours seem to arise frequently. This behavior can overwhelm an IPO. An IPO can control it more easily when it is the legal buyer.

Having the IPO issue orders to the supplier in the supplier's country gets around the issue of representatives or subsidiaries in the United States having exclusive sales privileges. The sale becomes a domestic sale in the supplier's country. This can save commissions and markups in the purchasing channel.

The IPO should only issue purchase orders on the authority of the actual buyer. There are two routes to give the instructions to the IPO. Some companies have a well-developed internal order system. The buyer issues an internal order to the IPO, who issues a local purchase order to the supplier. The IPO actually resells the goods to the buyer.

Other companies, with well-developed computer networks, are using the concept of a "remote buyer." An IPO buyer uses a computer network to enter the MRP of the purchasing location. The buyer receives suggested orders electronically and issues them to the supplier. This avoids doubling the administrative work, but requires trust in the integrity of the MRP process and the effectiveness of the remote buyer. The IPO becomes a remote extension of the domestic purchasing department.

### Payment

An IPO can pay the supplier in local currency and eliminate the need for letters of credit. It can also handle the details of exchanging currency and hedging, leaving buyers free to worry about purchasing issues.

### Logistics Management

The IPO can provide another very valuable service by managing the freight forwarding and shipping in the seller's country. They can research the best and cheapest routings, and can be sure that goods actually ship as the freight forwarder says they will. This eliminates a lot of doubt and un-certainty and reduces double-checking of the freight system.

## COSTS

Well-run, successful international procurement offices tend to operate at the expense level of 1 to 2 percent of shipment volume at a shipment level of more than $50M per year. Smaller IPOs, in the $5–10M per year range, cost approximately 5 percent to operate. IPOs need enough funding to perform the "shop the world" role of obtaining local quotes, and if there are multiple IPOs competing for the same business, by definition, not all of this work will result in business.

IPOs should charge users for their services. They need to develop a schedule of charges that are lower than what it would cost a buyer in the United States to perform the role. Use should be voluntary, so that the IPO concentrates on providing services that add value as perceived by the buyer.

The IPO needs to recognize that it is in competition with representa-tives, subsidiaries, and independent agents. It also has to overcome many buyers' preferences to perform the whole international procurement role without help.

## STAFFING

IPOs are primarily staffed with employees who are nationals of the IPO's country. Occasionally, it may be necessary to relocate a person from the buyer's country to get an IPO started, but usually the funding of such expatriates is too expensive for them to stay very long. The best employees would be found in an already-existing factory that the buying company has in the IPO's country. If the buying company has no such factory, an alternative is to transfer employees from a sales office.

I don't recommend setting up a stand-alone operation that will just be an IPO. This presents a very limited career path for a potential employee, and the best potential employees would be less likely to seek a job there. There would also be a significant legal overhead burden of incorporation, taxation, and the like.

## IPO LOCATIONS

I believe a well-run IPO is very valuable in circumstances where there are major cultural or language barriers between buyer and seller. An IPO is very useful in Mexico, for example. Major cultural differences can make both the buyer and the seller uncomfortable and less effective than necessary. However, none would normally be needed in English-speaking Canada, which is very close to the United States in culture and language.

IPOs can also be useful in any country if the supplier's processes need close monitoring. The culture and language of Britain are also very close to that of the United States. However, an IPO can still be useful in process industries where the supplier needs close control and surveillance.

## IPO WEAKNESSES

There are some roles that IPOs are less successful at performing. In many cases the buyer needs to perform the role directly.

### Design Help

Design issues and questions are best handled between the designers at the supplier and the engineers at the customer. The IPO should be aware of the issues being raised, but will not be able to add much to the process.

Therefore, the supplier's and buyer's engineers need to be able to communicate, at least in writing, in the same language (or the supplier must be willing to hire a technical translator). This will reduce the list of appropriate suppliers of any product except standard, off-the-shelf parts.

### Virtual Sourcing

Using an IPO does not eliminate all contact between the actual buyer and the supplier. The IPO is an agent, and an agent that will develop a reasonable degree of authority with the supplier, but it is not the ultimate supplier. The actual buyers need to become familiar with the actual sellers. If there is significant business, the buyers should request that the supplier visit them periodically.

## *Countertrade*

Countertrade refers to the practice of a company promising to buy material from a country in return for the privilege of selling there. Countries with noncompetitive industries impose these requirements. Some companies see an IPO as an easy, low-capital way to meet these requirements. My experience is that purchasing from such a country is more difficult than manufacturing there. A company that sets up a manufacturing plant can bring in larger resources to solve the local country problems than an IPO can.

If your company is setting up an IPO to solve countertrade or other trade barrier issues, it is absolutely critical that the purchasing department and the sales staff of the buyer's company work for the same profit center. This puts the benefits and costs under the same managerial responsibility. Both parties must have received clear directives and understand the benefits and costs of meeting countertrade goals.

## SHOULD YOUR COMPANY OPEN IPOs?

I'm a believer in IPOs. They offer many advantages at a relatively low cost. With e-mail systems, they are extremely easy to communicate with. They can actually work as your night shift. You send them a problem just before you leave for the day and you should expect a response when you return to work in the morning. You can get two day's work in one day this way.

They are the only way to have inexpensive, on-site representation if things start going wrong. An IPO should be able to get to a supplier in a few hours. This can save a great deal of money by avoiding the "one-more-shipment" syndrome.

You also get support for language, legal, and cultural issues. The value of this is hard to quantify, but it definitely makes travel and negotiations easier. Most negotiators welcome the support, but there are some who believe that they are capable of doing everything themselves with limited help. This may be true, or it may be a case of strong but misguided U.S. cultural individualism.

Finally, if the IPO places purchase orders, you get an easy method of paying in foreign currency. The IPO pays the supplier, and you pay the IPO through an internal payment system.

I believe that if your company plans to buy more than $10M to $20M per year from a region and has a successful track record of hiring and keeping good employees in that region, it should open an IPO there.

## *KEY POINTS*

▶ International procurement offices are good solutions to the problems of international purchasing for companies that buy more than $10–20M in a region and already have a subsidiary or joint venture in the suppliers' country.

▶ They should be staffed with engineers, buyers, and people skillful in marketing their services. Employees should be long-term employees, who know the needs and practices of the buying company.

▶ For best functioning, IPO use should be voluntary and IPOs should be required to generate revenues to cover their expenses.

*Chapter Twenty-One*

# Channel Management

The selection and management of a "procurement channel" are much larger issues for international purchasing than they are for domestic purchasing. A procurement channel is simply the route by which information, orders, and payment flow between a manufacturer and a customer. There can be several channels in use simultaneously, as orders can flow in one channel and payment in another, for example. The procurement channel is not the same as the logistics channel, which is the route by which goods flow.

There is a wide variety of channels available in international purchasing because there is a wide variety of organizations that assist buyers or sellers, but for a price. Looking at these organizations brings out the issue of the cost of their services versus the value they add.

Any organization between the actual buyer and seller adds costs. It may or may not add value in services that are worth more than the cost. In managing global supply, you need to work to uncover and unbundle these costs and then choose which ones to incur. This gets to be a game of strategy, as each intermediary will fight to maintain its position in the channel. A buyer can work to take maximum advantage of this situation.

Exhibit 21–1 shows the potential intermediaries (and potential channels). The intermediaries can be working for the buyer or for the seller, or they can be independent agents.

## SELLER'S SUBSIDIARY

The seller's subsidiary is the most common intermediary involved in big business transactions. These are companies such as Hitachi America or Siemens, Incorporated, that foreign suppliers have set up in the United States. These are generally sales organizations that make it easier for U.S. companies to buy from remote parent companies.

**EXHIBIT 21–1**
*Potential Channels for Orders and Communication*

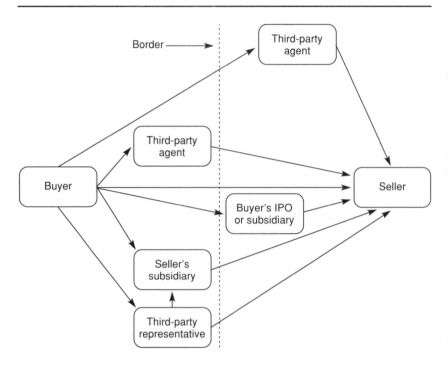

These subsidiaries have many apparent benefits. They are in the right time zones, they speak English, and they accept payment in dollars (which may add to your costs). However, they are not the supplier. If you are talking to someone in the subsidiary, that person is usually not a key decision maker. Consider how often you hear, "I don't know, I'll have to check with Tokyo (or Munich) tonight and get back to you." Then consider that you may be paying someone for something you could do yourself.

You need to open communication with the same level of decision maker at a foreign supplier as you do at a domestic one. Most major business arrangements and negotiations require participants to be above the level of sales representative. Your company usually needs to be able to talk to design engineers, and to product, manufacturing, and general managers. These people usually do not work at the subsidiary.

**EXHIBIT 21–2**
*Typical Large-Company Japanese Organization*

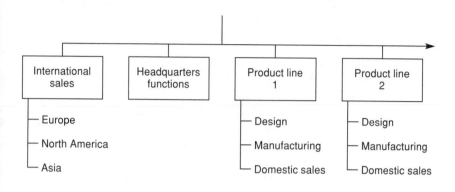

The subsidiary employees are geographically remote from the supplier's key management. They are often organizationally remote as well. Japanese companies that started as trading companies are usually organized with many international sales subsidiaries reporting to an "international division." This international sales group handles the sales of all the company's divisions. Domestic sales will belong to a product group, and will be tightly coupled to the product line management. Exhibit 21–2 shows a typical Japanese large-company organization.

The international sales connections to the decision makers can be tenuous. They are definitely the wrong choice to be the channel for performance feedback to the supplier. Your message has a good chance of being distorted or obscured, especially if it reflects badly on the subsidiary.

The costs you are paying to deal with these subsidiaries are some of the most difficult costs to discover. Selling companies who have set up these subsidiaries will prefer that you buy through them, and will make it difficult to uncover those costs. The costs I was able to determine ranged from 38 percent of a $7M annual purchase to 5 percent of a $40M annual purchase. Five percent doesn't sound large, but in this case it is $2M per year, which is several times the cost of a small IPO.

You need to consider whether the benefits of using the subsidiary justify these costs. If you can't determine the exact costs, use 5 percent as a minimum and expect 10–15 percent for business less than $10M per year.

For example, you might estimate that the subsidiary is costing you $500K–$750K per year on $5M worth of business. Is the subsidiary performing services that you could not do yourself for less money? Sometimes it does, but often it does not.

## BUYING DIRECTLY

This channel should result in the lowest cost of the purchased items, because there are no middle people in the channel. You will also get the most direct communication with the suppliers, provided that you can cross the language and cultural barriers successfully.

The tradeoff is in overhead costs. Your travel and communication costs will almost always be higher than when dealing with an IPO or with a supplier's subsidiary or representative in your own country. You will be completely on your own.

I suggest dealing directly only when you are completely comfortable dealing in the supplier's country and when the supplier has proved that its quality, delivery, and cost reduction processes are under control.

## INTERNATIONAL PROCUREMENT OFFICES

We have covered the basics of international procurement offices in an earlier chapter. IPOs can be considered to be either an intermediary or an extension of your domestic purchasing department, depending on your company's organization. Typically, if a company has multiple profit centers, the IPO will belong either to one of the profit centers or to a corporate organization. If the IPO is in a separate profit center than the buyer, then the IPO is closer to an intermediary. If it is in the same profit center, then it should act as an extension of the domestic purchasing department.

### Independent International Procurement Offices

Organizations are developing that are effectively becoming third-party IPOs. These companies act as your agent in a foreign country. They are not agents of the supplier. This is a worthwhile idea and such organizations are worth looking at. Here is a checklist I would use in evaluating them.

1. *Communications:* Can you communicate easily and reliably with the organization through your company's e-mail system? Most systems these days can send and receive messages to and from another company's e-mail system. Fax communication is a distant second best.

2. *Local Representation:* Does the organization have a sales and liaison office in your area? This will help a lot in their gaining understanding of your needs.

3. ***Integrity:*** You need to be sure that they are honest and ethical. You will quickly find out if they keep their word on issues of quality and delivery, but more subtle issues can occur. You need to have audit capability and contact directly with the suppliers so that you can be sure that you are being charged no more than an agreed-upon schedule. It would be easy for one of these organizations to neglect to pass on supplier cost reductions to you. Avoid organizations that try to prevent you from communicating directly with a supplier.

4. *Cost:* The cost charged as a markup must be reasonable. The independent IPO must understand this and must work to continue to drive down supplier's costs, even if this reduces their own revenue in the short term. They need to understand the worldwide competitiveness that your customers demand. Reducing a supplier's price gives a short-term penalty to a commission-based agent. You need an agent that thinks long term.

5. *Technical competency:* The independent IPO should have engineers able to evaluate a supplier's manufacturing process. It should also be able to operate a quality-assurance program. At a bare minimum, it must be able to inspect goods at the supplier's plant. It should have ability to audit a supplier's processes and QA system and to convince the supplier to improve systems and processes to a level you need.

## INDEPENDENT REPRESENTATIVES

Independent representatives of the supplier or of the supplier's subsidiary usually handle the 80 percent of a supplier's customers who result in 20 percent of the business.

Whether you must use these representatives or not depends on the size of your company and its purchases versus the supplier's total sales. Suppliers are legitimate in wanting to reserve the time of their managers and other key people for dealing with larger customers. A good rule of thumb is that if you would be dealing directly with a certain-size domestic company, you should be dealing directly with an equivalent-size foreign company. Size may be a criterion, but distance should not be.

You should be very reluctant to enter into arrangements where you deal with an independent representative of the U.S. subsidiary of the supplier. This gives you two layers of intermediaries, each one adding cost and delaying communication. If faced with this situation, try every alternative before accepting this channel. You might be better off dealing with an independent representative in the supplier's country.

Costs of independent representatives in the United States seem to vary with what the market will bear. I have seen costs as high as 72 percent on top of a $400K annual purchase, and as low as 6 percent on a $20M annual purchase. If you are going to buy something through an independent representative, be sure that the representative understands that you will want full cost disclosure of what the markup is. If a representative refuses to cooperate with you in this matter, you should escalate the issue to the supplier itself.

---

### Collisions in the Channel

Some of the most acrimonious negotiations I have been in were about channel selection. I did not expect this when I moved to Asia and started IPOs.

The most difficult circumstances were when the original quote for the business came from a sales subsidiary in the United States. If a buyer decided later to get an Ex works quote through the IPO, suppliers would frequently refuse. Those that did quote almost invariably came up with an Ex works price that "proved" that there was no benefit to buying Ex works. One supplier actually produced numbers showing it was more expensive to buy direct. He did this by unbundling his DDP (delivered duty paid) price in a way that resulted in an impossibly low price for shipment.

When you get quotes showing that there is no benefit to buying direct, you should recognize that there is still money left on the table. The supplier's costs will be reduced if business does not go through the subsidiary. With some strong negotiating, you will be able to save this money.

## BROKERS

Brokers are independent organizations who work to bring together a buyer and seller. They are paid a fee based on their business volume. Some buy and resell goods; others do not. Either the buyer or seller may contract with (and pay) them.

The main value that they add is the ability to locate sources for buyers and customers for sellers. They have a fundamental problem in their industry because they usually do not get paid for their services by a finder's fee, but by a commission based on business volume. This payment comes well after the major value they added, and this leads to attempts by buyers and sellers to remove them from the channel.

As a defense, some brokers will often try to keep the parties unknown to each other so that they cannot meet to take the broker out of the picture. It's impossible for a buyer to be successful without direct contact with the supplier, so you should avoid this type of broker.

If you do hire a broker, or work with a foreign supplier represented by a broker, be sure to discuss how long the broker's payment will go on. If he or she expects it to continue through the life of the relationship, he or she should add value throughout the relationship. Most cannot do that.

If you see value in paying a finder's fee, offer to do only that, and plan to deal directly with the supplier. If the broker refuses, look for another one. The maximum finder's fee you should pay is approximately what it would cost you to find the supplier yourself. This includes the cost of travel, translation support, financial analysis, and reference checking.

## TAKING CONTROL OF THE CHANNEL

If you decide to take control of the order channel, here are a few suggestions.

### *Establish Your Position Early*

For all new business relationships, state from the start that you, as the customer, reserve the right to choose the order channel. That identifies you as a cost-conscious buyer. Explain that as a part of any cost disclosure you will want to know subsidiary markups, costs of freight, and duties. This activity at the start will pay dividends later.

## Establish Relationships

If you are trying to move away from an existing representative or subsidiary relationship to more direct buying, you need to do some homework before you make your move. Someone at the supplier's headquarters will be involved in any decision to sell directly to you. Headquarters' first contact with you should not be your demand to deal directly. You need to take some time to create a personal relationship and to establish your company as one that they should be dealing directly with.

A good way to start to establish this relationship is by meeting the supplier for a performance review. These reviews should not be going to the subsidiary or representative anyway, so notify the subsidiary that you want to visit the supplier (or have the supplier visit you) for a performance feedback meeting. Use this meeting not only for performance feedback but to establish personal relationships. Be tactful if you must deliver a critical message, and take care to appear to be a reasonable company to deal with.

## Consider the Subsidiary's Role

You also need to decide what relationship, if any, you will have with the supplier's existing subsidiary or representative. You may decide that you need them to make technical sales calls to introduce new products to your designers. You could also decide that you can obtain enough new product information in periodic visits to the supplier's headquarters. Be sure that you know exactly what you want.

If you do decide that you need continued contact with the supplier's subsidiary or representative, the negotiation can become more complex. The representative or subsidiary will very likely tell you that they cannot call on you unless you place purchase orders through their office. Usually, this is a negotiation tactic only.

While the easiest and most traditional way to fund a sales subsidiary is through markups on orders, there are many other possible ways. The most common funding alternative is that the subsidiary will receive funding based on total purchases by your company. This should be less than their normal markup, because the sales subsidiary will not have to handle orders, returns, shipping, or logistics.

## *Make Your Move*

Once you know the key people in the supplier's plant, make your request of moving to direct buying to the supplier's sales management, at the supplier's headquarters. Have a convincing set of reasons.

The best reasons are the ones that you can put positively. You could state that your business volume is so large that you should be dealing directly. You could also state that the supplier is strategic to you.

You may also have some reasons that are more difficult to put positively. These might reflect poorly on the performance of the existing subsidiary or representative. Hold these reasons until you decide you absolutely need to use them. Remember, especially in Asia, that causing someone personal embarrassment is often more offensive than in the United States.

## *Be Persistent*

Do not expect to have your requests agreed to every time, but be persistent and you will eventually win most of them. It can take years to change away from a subsidiary-based relationship to a more direct relationship.

## *KEY POINTS*

▶ Intermediaries in a channel add cost. Try to determine what that cost is and whether the services the intermediary provides are worth the additional costs.

▶ A sales subsidiary of a foreign manufacturer is an intermediary, not the manufacturer itself. Effective supplier management requires contacts at the supplier. Expect a minimum cost of 5 percent for annual purchases of more than $10M for dealing with a sales subsidiary. Expect 10–15 percent for purchases between $1M and $10M.

▶ Moving to more direct purchases can be difficult because most suppliers will act to protect their sales subsidiary structure. A buyer must have contacts at the supplier's headquarters before attempting to move to direct purchasing.

*Chapter Twenty-Two*

# Supplier Selection:1

This is the first of three chapters describing a "procurement cycle" of searching for potential suppliers, narrowing the list to the most likely few, negotiating with the final candidates, signing a contract, and managing a remote relationship. The first task is to locate potential suppliers. While this often happens in reaction to a sudden need, it's better to take a strategic approach.

## TIMING

Many procurement departments write a "product procurement strategy." These strategies detail their company's needs for a product, the nature of the supply market, the expected supply base over the next few years, and the supplier management strategies that they will use.

These strategies are usually updated annually. This update is a good time to include a section on the global supply base. The section should describe:

- Countries producing the product.
- Names of potential suppliers.
- Plans for future activities to investigate and develop new sources.

Typically, this strategy section would look five years into the future.

## SEARCH PLAN

You should look for the best sources in the world. If you do not know much about potential suppliers, the first thing you should do is to identify the best countries. From there, you can move on to identifying suppliers.

## Country Development

Often, you will be investigating suppliers in low-cost, developing countries. Countries tend to develop in a structured way. First, a major multinational will move into a developing country and set up a factory. This factory will be mainly an assembly plant and will do very little local purchasing. The components will be imported.

Next, and with varying degrees of speed, supporting suppliers will grow up around the multinational and replace the imports. Eventually local employees will leave the multinational and start their own competing companies.

At this point, the country has what Michael Porter, in *The Competitive Advantage of Nations,* calls a "cluster," and the potential for world-class competitive industries. Countries that become excellent are also free of minor corruption, logistics difficulties, and other issues that can complicate international purchasing.

---

*Where in the World?*

Michael Porter's book helps to explain why it's difficult to use macroeconomic analysis of a country's resources and cost structure to predict which country will do well on a given product. I knew macroeconomics didn't work after an attempt to source power cords from Mexico. I reasoned that Mexico not only had low-cost labor, but was rich in petrochemicals and copper, which are the materials used in power cords.

At that time, power cords from Singapore cost about $1.00 Ex works, and cords from the United States about $1.80. The first prices from Mexico were a disappointment, with costs of about $1.40. Porter's book explained the missing ingredient, which was a base of demanding customers. Mexican power cord manufacturers had learned that they could price their cords 10 to 20 percent below U.S. prices, and get the business of customers who were not shopping globally. The demanding customers went elsewhere, including Singapore, which had high labor costs and no natural resources.

It took three years before some Mexican suppliers saw the need to compete on a global basis, and now Mexican cords are cost competitive.

---

There are distinct stages that interest a buyer. When a major multinational moves into a less-developed country, there is a good chance that the multinational may be able to offer lower prices quickly. If you are buying from the company already, you should know and explore these possibilities.

Later, when the supporting industry starts to develop, it becomes attractive to buyers of the supporting products. You do not want to be so early into that market that the supporting company is captive to the multinational, but you should start checking early and be prepared to move to be the supporting company's third or fourth customer. The major multinational that the supporting companies supply may be your competitor, and you will be looking at your competitor's suppliers.

Finally, when local competition to the multinational starts, you will want to look at those suppliers.

## COUNTRY INFORMATION

Information on which countries are successfully building a product is easy to obtain. Simple reading of industry magazines covering the products you are purchasing will give you this information. U.S. magazines will occasionally have articles talking about foreign competition, but it is better to subscribe to foreign magazines covering your industry.

If you read a foreign language, this gives you an advantage. However, there are English-speaking countries around the world. Besides Britain, India, Singapore, and Hong Kong all have magazines published in English. You should search them out and subscribe. I have listed a few in the "resources" section of this chapter. If foreign magazines do not cover your industry, you should first uncover the most likely countries through reading U.S. literature. Next, contact the trade development offices of appropriate countries to inquire about magazines and newsletters.

---

*Country Development Examples*

Singapore is a good example of the country development process. It long had a low-tech semiconductor assembly operation, but this was largely either major multinationals or small local subcontractors. Very little could spin off from this industry. There was some development of local plastic molding and tool-making industry, but the industry itself did not develop a "cluster" and did not remain competitive.

*(Continued)*

---

*Country Development Examples* *(Concluded)*

However, multinational household TV and electronic appliance companies from the United States and Japan started moving to Singapore. Computer companies followed. Many of these companies started subcontracting their printed circuit assembly work, and a printed circuit assembly industry developed. This industry remains competitive. In addition, higher-level assembly houses opened under local management, and now many of the multinationals in Singapore subcontract their work to these local companies. In turn, these assembly companies are selling their services directly to foreign companies.

Sometimes this works in another direction. Britain has long had an excellent precision metalworking industry. It also has recently developed a competitive electronic assembly industry. It has all the potential for developing a disk drive industry, as disk drives are largely electromechanical parts with some electronics. While this has not developed yet, I predict it will.

---

## NATIONAL TRADE DATA BANK

One approach for a company just starting on importing is to see what other companies are doing. The U.S. Department of Commerce keeps detailed statistics on imports. They maintain a database called the National Trade Data Bank. It shows imports by country, sorted by harmonized system (customs) code. If you know the HS code of the product you are researching, you can find which countries are successfully exporting those products to the United States.

The statistics cover a three-year period, so you can look at trends. If, for example, Japan has high but declining exports and Taiwan has lower but increasing exports, you should not overlook Taiwan.

The list of countries that you develop will be a starting point. It will probably not identify very many suppliers yet. This list of countries will also have a short shelf life. Countries' competitiveness changes rapidly, especially as currencies change. You should review this list annually, as part of your strategy update.

---

### Political Stability

People often ask where political stability fits in when deciding where to purchase. I think this question is overemphasized and is often used as a reason to continue with current sourcing practices. Exports are usually the last area to be impacted by political instability, especially in developing countries.

For example, Intel operates a wafer-fabrication operation in Israel that continued to operate uninterrupted during the uprising of the last eight years. This plant is not in the presumably "safe" area of coastal Israel. It's in Jerusalem, right in the center of activity. A few years ago, it was one of the few plants producing the critical 80286 microprocessor, and there was never any evidence of an interruption.

Similarly, Texas Instruments operated a key plant producing a crucial family of digital integrated circuits in El Salvador through the revolutionary activity of the early 1980s. This plant also produced without interruption.

Thailand has had several coups in the last few years. General reports are that government-run facilities, such as ports, actually improve their efficiency after coups.

There are newsletters about political instability, some of which increase their sales by needlessly scaring people. A good indicator is to watch capital investment or divestment, and newsletters focusing on this issue may be worthwhile. A good deal of sophistication is needed, however. Did any newsletter predict that Slovenia would remain stable and Bosnia become unstable when Yugoslavia dissolved?

---

## POTENTIAL SUPPLIERS

Once you have determined countries that you want to investigate, the next step is to develop a list of potential suppliers. At this point, you should make a minimal investment in time and expense. You should also minimize the supplier's time and investment in providing information. The goal is to move quickly to a relatively small list of suppliers. Unless you have an IPO or a good manufacturing subsidiary in the country, you should expect to make at least one international trip in this process. If you have searched countries that have a successful track record in the product you are seeking, you are more likely to have too many suppliers than too few.

The March 1995 edition of *Electronic Components* has advertisements from 25 computer cable assembly companies in southern Asia, for example. Picking the best few of these suppliers can be a challenge.

You also have a potential to overlook a few key suppliers unless you use more than one method to obtain supplier information. For example, *Electronic Components,* one of the best sources of information, did not have an advertisement from the major Japanese multinational that is one of the region's key suppliers of computer cable assemblies.

However, you need to guard against an expectation that you will automatically be able to zero in on the best supplier in the world the first time you try a global search. If you get to the right country and choose a supplier who is a major improvement over your present supplier, you will eventually hear of even better suppliers through the contacts you develop in that country. You will still have improved your company's situation.

## COMPANY INFORMATION SOURCES

There are a lot of sources of information. These sources will vary in the breadth and depth of the information. Some sources will be able to give you very complete information on a few companies. Others will be able to give you limited information on a lot of companies. Here is a list of a few sources.

### Trade Magazines

These magazines will give you reasonably broad information on supplier names, addresses, and fax numbers. This can be either in advertisements or in articles. Generally, though, these magazines are better at aiming you at the right countries than at the right suppliers.

### Country Trade Information Offices

Almost all countries maintain one of these offices, whose role is to develop their country's exports. They are usually in their embassy in Washington, D.C., or their consulate in another large American city, such as New York,

Los Angeles, Chicago, or San Francisco. Check the phone book under "Consulates." Taiwan is an exception, as it removed all governmental bodies from the United States when the United States recognized the People's Republic of China. There is a "private" organization called the Taiwan Economic and Cultural Office (TECO) that provides the normal consular and trade development roles, such as issuing visas and developing trade. They have offices in several U.S. cities.

Information from these organizations will be broad. Because they are tax-supported organizations, they may not be comfortable in trying to make value judgments about individual companies in their countries. They should be willing to give information on size, however.

The following information sources are in the country you will be investigating. You will have to make at least one trip. If you have a travel budget that can afford two trips to the country in question, I recommend that on the first trip you only meet these in-country information sources, and not visit suppliers.

### *Your Offshore Subsidiaries*

If you have a factory in the country you are considering, you should ask that factory for help. If you only have a sales office or representative, they may be of some help also. However, a sales office may not understand your manufacturing needs and might steer you toward their customers. At the very least, either a factory or a sales office can assist you in finding valuable information resources in their country.

### *U.S. Chamber of Commerce*

Most countries will have a U.S. Chamber of Commerce (sometimes called American Chamber of Commerce). This organization helps to provide services to American companies in the area. If you contact them ahead of the trip, you can see if they have full-time employees who can help you and recommend other people for you to contact. If this is your preliminary fact-finding trip, the Chamber of Commerce can provide names of people at other U.S. companies for you to talk to. They may be able to provide you with information about business practices and cultural and linguistic hints about doing business in the country.

## *Other U.S. Companies*

Other U.S. companies operating in the area can be excellent information sources. If you can obtain names of your U.S. customers or existing suppliers who have factories there, you may be able to find out useful information from them. I have even seen competitors talking. An approach of "from one American to another" seems to work well. The ideal company to talk to is one that operates an IPO in the country. They should know the companies who are exporting successfully.

## INITIAL CONTACT

You are now ready to make initial contact with your list of suppliers. Fax is an excellent tool for this. E-mail is even better, but is often not available. You can prepare a cover letter that describes:

1. *Your company*  Sales volume, locations, products built, years in business. If you have a prepared press package, save this and send it only if the potential supplier asks.

2. *Yourself*  Name, title, position, and how to reach you by phone and fax. Be open to phone calls, and talk to anyone who calls. They will be checking you out also. Remember the importance of a personal relationship.

3. *Your project*  You are uncovering preliminary information on suppliers of a particular product. Give a REALISTIC estimate of annual unit volume. Avoid exaggerating your purchases.

4. *Request for information*  Ask for catalogs or data sheets on products they sell, annual sales volume, sales contact name and phone numbers, and multinational references. This is the information needed to eliminate companies who have no experience selling to foreign customers, whose sales are too small, or who cannot communicate adequately in written English. You should also take care to be sure that you have reached a manufacturer, not a broker or other intermediary.

If you have decided that you are not going to use representatives, you should also inform the company that you do not want a call from any U.S. representative or subsidiary at this time.

I do not recommend getting any price quotes at this point. You will be suffering from information overload and there is no point in wasting the time of unlikely suppliers. You can make an exception if you are researching a standard, easily identifiable commercial item, or if you need a price quote to justify a trip.

## ELIMINATING SUPPLIERS

You should now start reducing the supplier list. You should eliminate suppliers who are too small, too clumsy in their response, or who have no multinational references. (Unless you have an IPO, I don't recommend being the first export customer of a supplier.) If this still leaves you with too many suppliers, start checking references. If the list is still too long, ask for financial statements next. (The next chapter gives some warnings about use of foreign financial statements.) The goal is to reduce the supplier list to a length that you or your IPO can afford to visit. A good rule of thumb is to limit the list to no more suppliers than you could visit in a week.

If you do eliminate a supplier, be courteous and tell the supplier that he or she will not be considered this year. You might or might not choose to add that you will keep the supplier in mind for future business.

You should now have reduced your list of potential suppliers to no more than 5 or 10. They should be in no more than two or three countries. Your next step is to reduce the list to two or three of the best suppliers. Following this step, there will be a last discussion with the finalists, followed by a decision.

## PRICE QUOTATIONS

Now you should obtain price quotations. You should be sure to get Ex works quotes. If the products are cost driven in pricing structure and are from a country with a floating currency, you should ask for quotations in the currency of manufacture. This should get you the lowest or base price of buying the part. You should add the costs of freight, duty, and inventory to develop a total "landed cost." The concept of unbundling the costs is developed as part of *Zero Base™ Pricing,* an excellent resource.

If you have concerns about your own organization's ability to handle freight effectively, you might also ask the supplier for a freight quote to the U.S. port of entry. This will result in a total quotation package that is effectively Delivered duty unpaid (for air) or Delivered ex-ship duty unpaid (for ocean freight). The only circumstances under which you should consider a quotation "Delivered duty paid" would be if there is no duty on the parts.

You should ask the suppliers to quote on realistic quantities. You should also expect to talk with the suppliers and allow more than one attempt at quotations if there are misunderstandings. During this process you will not just be getting price information; you will be determining how easy the supplier is to work with.

## FINANCIAL DATA

If you have not already done so, now is the time to ask the supplier for detailed financial data. You need enough data to determine whether the supplier is financially sound. However, be careful about financial statements. Financial terms mean different things in different countries, and the standards vary. Attempting to use rigid U.S. financial models in emerging countries may lead you either to reject a supplier needlessly or to accept one who truly is in trouble.

You should contact the office of a major U.S. accounting firm in the supplier's country. This firm can advise you on appropriate local standards. Your company's sales organization in the supplier's country can be another resource. They will have standards to determine whether to issue credit to potential customers. These standards can be also be applied to potential suppliers.

## REFERENCE CHECKING

This is also the time to check references. Be persistent and critical, and really zero in on any uncertainty and vagueness that references supply. Ask the reference about the items that are crucial to your company. Has the supplier met commitments? Is lead time under control? Do they ship on time? How is their quality? Has the supplier acted responsibly with respect to intellectual property?

Depending on your priorities, you can also reference-check after an on-site survey.

## SUPPLIER VISITS

I believe a visit to the supplier is essential in all but the most routine situations. If you would not make a strategic supplier change to a new domestic supplier without visiting the plant, you also should not change to a foreign supplier without a visit. If your company has an IPO, they will be able to save some travel money by making the visit on your behalf. There is no substitute for your company making the personal contacts and developing the perceptions that result from face-to-face meetings.

You can make the most effective use of your trip if you keep the following simple guidelines in mind:

- Your travelers should be stable in their jobs. Someone who is about to change jobs should not make the trip. It's better for you to present an image of continuity to the supplier.
- Travelers must do some basic reading on the country's history, culture, and language before the trip.
- If you are going to obtain assistance from someone in the supplier's country, be sure that person knows ahead of time what you expect. Plan to arrive a day ahead of time to meet that person and review roles and plans for the visits.
- Bring lots of information on your company and your place in it.
- Check ahead of time to see if you should bring translated business cards. You should translate cards for use in countries where the Roman alphabet is not used. Be sure to bring the right language cards. I have seen Japanese cards used in South Korea, which is equivalent to using a German card in Holland or France.

## SURVEYS

Usually a survey is taken when visiting a supplier. Surveys are a frequently overused tool, but they are appropriate if they are done in person, either by the buyer or by a carefully prepared representative who works for your company. They should be a tool for developing communication and understanding, not a sterile exercise in filling out pieces of paper.

Communication will develop when the supplier asks, or you explain, why you are asking certain questions. For example, if you ask for the on-time shipment rate for last year, this year, and the goal for next year, you and the supplier will likely start to talk about the importance of on-time delivery. You can find out whether the supplier agrees that your goals are reasonable and wants to meet them.

If more than one group fills out survey forms (for example, one team goes to Europe and one to Asia) it is important that the survey forms not ask for value judgments, but "just the facts." You can find out more by asking for the value of work-in-process as a percentage of weekly shipments than you can by asking if the production process appears to be well planned and flowing smoothly.

You need to develop knowledge of four significant facts:

- How will the supplier build and test your product?
- What level of quality will you get from this supplier?
- What percentage of the shipments will be on time?
- Will the supplier be staying in business?

To determine this, I recommend focusing on four key areas while surveying a plant. There is a suggested survey format in Appendix B at the back of this book, which may be copied or modified.

### General Information

Some of the basic information you need to know has to do with the size of the company, company management, and company stability.

**Size of company.**   You want to be sure that the company is large enough to meet your volume demand, but small enough that you will be regarded as an important customer. I like to account for between 5 and 10 percent of a company's sales. You can ask questions about employee count and sales volume.

I also like to ask about growth plans. Several questions in the survey ask for data from the previous year, the current year, and plans for the following year. This will help you to understand how carefully management is running the business and how carefully they plan for the future.

**Ownership and management.**   You will want to know the form of ownership of the company, who the key owners are, and who the key management is. You should check that key managers can communicate with you, at least in written English.

**Customer concentration.**   This is an issue that is related to company size. If a supplier has one customer who accounts for more than 30 to 40 percent of the company's sales, be careful. If that customer were to go elsewhere, the survival of your supplier would be at stake.

You should also check that you will not be the supplier's first foreign customer, and get the names of other foreign customers that you can use for reference checking.

**Financial data.**   The data in the survey will be sufficient for most financial analysis. It has balance sheet information such as the amount of debt and the amount of annual interest. It also shows operating statement information, such as sales and profits. It asks for three years' information, so you can look at trends. Remember to use analysis techniques suitable for the supplier's country.

---

*Financial Analysis*

One of the major factors that can throw off financial analysis is the difference in the banking system between many countries and the United States. Taiwan, for example, has a number of smaller banks that have family ties to whole groups of manufacturers. Just as an experiment, I once ran a standard U.S. financial stability check on the two largest electronics companies in Taiwan. They both scored as approaching bankruptcy. They are both still in business eight years later, and thriving.

The model did not take into account the close relationship the suppliers had with their bankers, who were very likely members of the same family.

The bill payment cycle is another area of country-to-country differences. In some countries bills are paid in 10 days. In other countries, including some developed European countries, the norm is 90 days. The large accounts receivable and payable balances that result can make an unprepared analyst nervous.

---

**Labor stability.** Find out whether there has been labor unrest or strikes in the recent past. (Some countries may not allow strikes, but there are fewer and fewer of such countries.) You should also get the name of the union, if any, that represents the employees. If unions are not required in the supplier's country, you should inquire why this particular company has one.

## *Manufacturing Practices*

You will want to know how the supplier will produce your parts, whether they will be shipped on time, and the supplier's quality practices.

**Production process.** You should look at the equipment and production line that the supplier proposes to use to produce your product. How old is it? Is it automated? Is there statistical process control (SPC) in use?

Here's where a cultural difference may arise. I believe SPC is always good and often essential. In the United States, we like to see machine operators making the measurements and plotting the charts. However, in countries with higher power distance (more hierarchical countries), it may not fit the culture to work that way. You might have to settle for trained specialists doing any SPC measurements.

**Work instructions.** How are work instructions conveyed to production workers? When feasible, I like to see written instructions, drawings, and/or samples at each worker's station. I avoid situations where workers are trained off-line and then are expected to remember the instructions for a long period.

**Subcontracted processes.** Pay careful attention to any manufacturing steps that the supplier will not do itself. You should find out what subcontractors will be used. If the processes are critical to success, you should also visit the subcontractor.

**Work in process.** You should ask for data on the amount of work-in-process inventory on the production floor, in weeks or days of production. Compare that number to the theoretical minimum cycle time that would be achievable if the parts were constantly being worked on. The closer the ratio of these numbers is to one, the more organized, flexible, and efficient the manufacturing process is. If there is more than two or three times the manufacturing cycle time of production on the floor, this is a sign that the supplier is going to be relatively inflexible and is probably having production problems.

**Manufacturing cost reductions.**   You should ask questions about cost trends in manufacturing. Total costs may historically go up or down in your industry, and you should look for appropriate attention to the issue. Even in industries where total costs tend to go upward, there are always ways to reduce individual costs. A good supplier will be paying attention to this issue.

### Materials Management

All buyers know how important materials management is to successful manufacturing. Purchased materials usually account for the majority of manufacturing cost, and supply chain management is the key to maintaining flexibility. Foreign suppliers are no exception. Here are some key items to check when evaluating a supplier's materials management.

**Order drivers.**   What causes the purchasing department to order parts for you? Do they respond to forecasts of sales and shipments, or do they wait for a firm order? A supplier who will not order material until there is a firm order will be less flexible than one who orders against a forecast.

This is an area for useful dialogue. You should see if you can improve delivery flexibility by sharing forecasts, or even by contractually obligating yourself to ordering within a set percentage of your forecast. This may get you the flexibility you need.

**Supplier lead-time control.**   Control of lead time from suppliers is the most important aspect of a supplier's being able to control lead time to you. You need to explore how the potential supplier will control its suppliers' lead time. Generally, the most effective way is through sharing of forecasts and open communication. However, in some countries, there can be almost a parent–child relationship between a company and its suppliers, and it would simply be unheard-of for a supplier not to respond as demanded of it.

**Imported materials.**   Does your potential supplier have to import materials? If so, do they seem sophisticated enough to manage both the process and the remote supplier? Knowing the imported materials and their sources can get you valuable information for negotiating and structuring prices. If a large part of, for example, a Taiwanese supplier's material comes from Japan, you will need to explore the issues of price protection and hedging.

**Change notice requirements.**   Most U.S. companies require suppliers to notify them if there will be any changes in purchased parts that could possibly affect applications or reliability. You should be sure that your potential supplier imposes similar requirements on its suppliers. Without this requirement, you will never be certain of design stability.

**Supplier management techniques.**   The most important issue here is whether your potential supplier is managing its suppliers at all. Ideally, a supplier will be actually generating data on delivery, quality, cost, and flexibility. The supplier will use this data in ranking suppliers, determining market share, and making decisions on future suppliers.

This is another area where there will be cultural differences in technique. It is very worthwhile to learn what techniques suppliers in other countries use. Some of these techniques might be very beneficial to your own company.

## Quality Management

Quality is the last, but the most important, factor on this list. Without good quality, the lowest-cost supplier in the world will not be acceptable. Here are some items to look for.

**ISO 9001, 9002, 9003.**   The ISO series of quality standards is becoming more and more a requirement worldwide. However, it is a quality systems and documentation specification, not a quality requirement. A company produces a set of operating specifications that guarantee consistency in its processes and its production. If the operating specifications meet ISO requirements and the company passes an audit showing it follows the specifications, it can obtain an ISO certification. This reduces, but does not eliminate, the need to survey a quality system.

There are two problems to consider. First, the specification guarantees consistency, not quality. The specification could state that a defined level of quality that you regard as mediocre would be acceptable. If the potential supplier follows the operating specification consistently, they will produce to that quality level consistently. It may not be good enough for you. Second, once granted an ISO certification, the supplier may not follow the specification consistently. There can be years between auditors' visits.

For these reasons, even an ISO-qualified supplier needs some quality auditing. You should check the quality level produced. You should also look for any practices that seem less than ideal and ask to see the ISO operating specification for the processes involved. You can then see if the ISO specification is being followed.

**Quality level.**    You should ask what a supplier's outgoing quality level is. What you are really looking for here is the attitude toward shipping faulty parts. There are still suppliers who ship to acceptable quality levels, or AQLs. An AQL of 0.4, for example, means that a lot with 4,000 parts per million (0.4 percent) faulty parts has a 95 percent chance of being shipped. This AQL is measured by sample inspection, and there are sample plans that allow a lot with known rejects in it to ship.

Unless your company is familiar and comfortable with handling faulty parts, you should steer clear of any supplier who proposes to ship lots that are known to contain faulty parts. It's generally cheaper to find a supplier who will clean up manufacturing processes and not produce scrap.

**Quality planning.**    Another key issue is whether quality improvements are being planned or just happen for specific customers who ask for special treatment. If you ask what the outgoing quality was last year, this year, and will be next year, you will learn the supplier's attitude toward quality planning. A good quality manual will also give you an understanding of the supplier's attitude toward planning and carrying out quality improvements.

**Calibration.**    You should check whether measurement equipment has a calibration label, and whether the labels are up to date. You should also investigate traceability of standards. There should be traceability at least to the potential supplier's own government standards bureau.

**Reliability testing techniques.**    If you do reliability testing on the supplier's product, you should ask the supplier to perform the same, or equivalent, tests. Once confidence has been developed in the supplier's integrity, you can often stop testing and simply ask for the supplier's data.

At the very least, there should be some sign of reliability information on the supplier's products. This might be from supplier tests or from other customers' tests.

## NEXT STEPS

After the surveys, the reference checks, and price quotations, you should be down to a small handful of potential suppliers. The next chapter will describe more screening steps.

## *KEY POINTS*

▶ Supplier search is best done periodically, as part of writing or updating a product procurement strategy.

▶ The best suppliers of particular products tend to group into a few countries that specialize in building that product. Discovering the best countries can shortcut the process of locating the best suppliers. Exceptions exist, usually when multinational companies open assembly plants in a developing country.

▶ Once a few countries have been selected, searching for the best suppliers is best done by searching out American contacts in the supplier's country. Other techniques are also possible, such as requesting information from country consulates or trade offices, becoming involved in an electronic supplier search, or simply responding to magazine advertisements.

▶ Selection of a supplier of strategic parts or services requires travel. It also requires not just knowing facts about a supplier, but developing understanding between the supplier and the purchasing company.

▶ Surveys should be done only in person or through an IPO. If done right, they are complex and difficult, and should be reserved for the likely few suppliers.

# RESOURCES AND REFERENCES

Asian Sources Media Group publishes *Electronic Components* and a number of other magazines that focus on Asian manufacturers of electronics, hardware, consumer products, fashion, and watches. They are based in Hong Kong, and have a U.S. liaison office in the Chicago area. Phone number is (708) 475-1900.

Burt, David K.; Warren Norquist; and Jimmy Anklesaria. *Zero Base™ Pricing.* Chicago, IL: Probus Publishing Company, 1990. (Zero Base pricing is a trademark of the Polaroid Corporation.) An excellent book on price negotiation and cost reduction strategies.

Global Trade Information Services (GTI) is an international trade and marketing consulting company. It works with governments and corporations to help better access the U.S. market. It publishes U.S. and Japanese trade data on CD-ROM, in a series called the World Trade Atlas™. They are at 2218 Devine Street, Columbia SC 29025, and can be reached by phone at (803) 765-1860.

Morgan Grampian PLC, Royal Sovereign House, 40 Beresford Street, London SE18 6BQ, UK, publishes and distributes several trade journals. Their telephone is (+44) 81-855-7777.

National Trade Data Bank; Available in CD-ROM from the U.S. Department of Commerce in Washington, DC. An on-line query program is available by subscription through the Internet.

Porter, Michael. *The Competitive Advantage of Nations.* New York: The Free Press, 1990, Chs. 3 and 4. This book gives an excellent and detailed view of how world-class industries develop and cluster.

## Chapter Twenty-Three

# Supplier Selection: 2

After the survey, you should eliminate all but three or four potential suppliers. You might eliminate them for reasons of quality, obvious high costs, or simple lack of compatibility in goals and philosophy. You are now in a position where you have to do an accurate job of determining the cost of buying from potential suppliers who have different currencies, duties, and shipping costs. The best way to determine this price is by calculating a "landed cost."

## LANDED COST

A landed cost is the total cost to buy the products, bring them to your receiving dock (land them), and finance the inventory required both in your stockroom and in the shipping pipeline. This includes:

- The cost of the part.
- Shipping costs.
- Duties, broker fees, and customs fees.

If you are considering a new part or a new supplier, you should add the cost of any tooling (don't forget assist duties) and qualification visits. This is a simple exercise when you are comparing one supplier with another and have only one buying location. Spreadsheet software is convenient for this exercise.

### Data Needed

To calculate a landed cost, you will need to assemble the following data:

**Company-specific data.** You need to know your own company's accounting cost for holding inventory. This tends to run between 12 and 25 percent per year. This cost should be applied to

- Your normal stockroom inventory.
- Any extra safety stock required because of the supplier's distance.
- The stock in the pipeline between you and the supplier if you are taking possession Ex works.

**Project-specific data.** "Project" refers to the use of the purchased part. It is generally a product or a family of products that your company sells. The data needed is

- The average volume in units per week (or other time unit).
- The life of the project. You need this data to amortize tooling costs.
- The percentage of the finished product that you will re-export. You will use this to calculate any duty drawback that you are entitled to.

**Part-specific data.** You need to know the weight and the volume (physical size) of the product you are purchasing to calculate shipping costs.

**Supplier-specific data.** This is the data that varies from supplier to supplier. It includes

- The cost of the part.
- The currency or currencies that the part is priced in.
- The exchange rate of those currencies.
- The duty rate for that part from the supplier's country.
- The cost of your customs broker to clear a shipment.
- Freight costs for a 20- and 40-foot ocean freight container, as well as for air freight. This information, along with the transit times, can be used to determine the optimum shipment method and frequency.
- Transit times for ocean and air.

- Safety stock required for this supplier, if any.
- Cost of tooling, if any, and the currencies and exchange rates involved.
- Cost to qualify the supplier, if any, in U.S. dollars.

### *Landed-Cost Calculation*

There are many ways to calculate a landed cost. You need to treat all suppliers equally and focus on the areas that clearly differ from supplier to supplier. Freight costs will clearly differ from supplier to supplier, for example, so freight costs are part of the landed cost. Cost of poor quality does not clearly vary from supplier to supplier, so it should be excluded from the landed-cost calculation. You should eliminate a supplier with a significant chance of poor quality from consideration prior to this step.

It is difficult to tell in advance how much of a buyer's time or travel money a particular supplier will need, so these costs can be put in a "risk" category that will be described later.

**Part cost.**   This is the cost of the part in dollars. If it was not quoted in dollars, you need to divide the foreign currency costs by the current exchange rate to convert to dollar cost.

**Tooling costs.**   Convert the tooling costs to dollars using the current exchange rate. Add to this the duty on the tooling (remember, it is an assist) even if the tooling never enters the United States. The duty rate on the tooling is the same as the duty on the parts. Divide this total by the number of units that the tool will produce. This gives you the tooling cost per unit.

**Qualification costs.**   Divide the qualification costs by the total units you will purchase. This will give you the qualification costs per unit.

**Freight costs.**   You need to choose a freight method and a shipping frequency for each supplier. If you used the freight optimization formula given in a previous chapter, this will give you a suggestion. However, you need to apply judgment in determining how many units can safely fit into an ocean freight container. You may not be able to fill a container completely if the items and their packaging are not strong enough.

If you have decided to ship by ocean freight and know the frequency and container size, you can determine the number of units per container. Divide this into the sum of the cost to ship a container plus the customs broker charge. You will then have the international shipping cost per unit. Next, add the unit shipping cost from the port of entry to your dock.

If you have decided to ship by air, and you have any reasonable volume of shipments, the air freight cost per unit will not vary much with the shipping frequency. Some costs will vary on a per-unit basis, such as the cost of the customs broker, but these costs are generally minor. You can optimize your inventory costs by shipping as frequently as possible, up to the point where unit shipping costs start rising because your shipments are too small.

If you do use air freight, multiply the "weight" (actual or volumetric) of a total shipment times the freight rate. Add the customs broker's charges to enter the shipment. Divide by the total number of units in a shipment. This gives you your unit air freight cost.

**Duties.** Add the customs charge from the Customs Service (currently 0.22 percent) to the duty rate to get a total percentage charge. If you're shipping by ocean, add the 0.125 percent Harbor Maintenance fee.

Multiply this percentage by the cost of the part in dollars. This gives you the duty cost per unit.

If you export some or all of your finished product, and expect to claim duty drawback, deduct the duties on the product you export. Do this by multiplying the duty rate times the percent exported times the unit price in dollars and subtracting this value from the unit duty. Remember that the Customs Service fee and Harbor Maintenance fee are not refundable.

**Inventory.**   You need to determine an average number of units in inventory, considering three types of inventory:

- Calculate average transit inventory by multiplying the transit time in weeks or days times the number of units used per week or day. This is true regardless of shipping frequency.
- If you use inventory on a steady schedule, then average variable stockroom inventory is one-half of the units delivered per shipment. If you use 1,000 units per day and a shipment arrives every five days, there are 5,000 units in the shipment. Assuming linear production, your average inventory is 2,500 units. If you

produce on a different schedule, then your average inventory will be different.

- Safety stock inventory is constant and assumed never to be used. Use the number of units of safety stock as the average safety stock inventory.

Add the three kinds of inventory, and you now have the average number of units of inventory. Multiply this by the unit cost in dollars, and multiply this answer by the percentage inventory holding cost per year. This gives you the inventory cost per year. Divide this by the number of units used per year to get the inventory cost per unit.

**Total cost.**   Add the part, tooling, qualification, freight, duty, and inventory costs to determine the total landed cost. You might or might not choose to buy from the lowest total cost supplier, but first I recommend some risk analysis.


## RISK ANALYSIS

Sometimes the supplier with the lowest landed cost may not be the best choice because of various risks. Some of the risks you need to consider are:

- *Exchange-rate changes*   The U.S. dollar could weaken against the supplier's currency. As a result the lowest-cost supplier may turn out to be more expensive than the others at some time during the contract's life.

- *Premium freight requirements*   You should expect your supplier to pay for any extra freight costs that result from its own inability to meet committed shipment dates. However, you may require expedited freight due to your own schedule uncertainties. A sudden increase in sales may result in major unexpected air shipments.

- *Pipeline scrap or rework*   The supplier should pay for any rework or scrap required due to quality problems that are the supplier's own fault. However, your company may have sudden, unexpected engineering change requirements, especially at the start of the production process, that could cause scrap or rework.

- *Extra procurement overhead expense*   You may need to spend
  more time and money traveling and communicating with some
  suppliers than others. This may be due to distance or supplier
  competency issues.

I'm not considering political risk at this point. Areas deemed too risky
should have been eliminated from consideration earlier than this step.

### Risk Analysis Methods

The four risks are easily quantified. Simply compare the total cost of
buying from the lowest-cost supplier to the other suppliers, foreign or
domestic. Calculate the total savings over the period of the contract, and
then:

- Determine how much the supplier's currency would have to
  change relative to another supplier's currency before the lowest-
  cost supplier is no longer the lowest. Check the historic variation
  of the supplier's currency and see if you are likely to have a
  problem. Consider this in light of your ability to hedge
  currencies and to find another source for the product.
- Calculate the extra cost to ship by premium freight, and
  determine what percentage of the parts you would have to ship
  that way before the lowest-cost supplier is no longer the lowest.
  If you planned to ship by surface freight, air freight will be the
  premium method. If you already planned to ship by air, this risk
  is minimal, as air freight costs vary very little by volume. Your
  major expense will be the increased customs broker fees.
- Take the amount of savings, and divide by your company's
  average cost of procurement personnel. This gives you the
  amount of full-time additional people who would have to work
  on this supplier to make it successful. Check to see if this is an
  amount you could possibly spend.
- Determine the average total inventory and see how many times
  you would have to either scrap or rework it before total costs are
  higher than the next lowest-cost supplier. Estimate, based on
  your knowledge of your company's design procedures, how
  likely this is to happen.

## ELIMINATING SUPPLIERS

Analyzing the landed costs and risks should enable you to reduce your potential supplier list to two or three suppliers. If the suppliers score approximately equally, you may need to contact more references and obtain more information. If this cannot be done, you should keep the suppliers you are most familiar with and most comfortable with, regardless of distance. You are now ready for a last round of negotiation and a final decision.

## FINAL NEGOTIATION

There are two reasons for this final negotiation. First, suppliers in some countries may expect it. There may still be flexibility left in the prices, and you will be leaving money behind if you do not reopen price questions. However, suppliers in other countries may consider the tactic devious and you might cause damage to your relationship. You need some country-specific cultural advice on this issue. Generally, southern Asian cultures will accept a final negotiation on price, and northern Europeans will not.

Second, you need to be sure that the remaining potential suppliers understand the details of the business arrangement. It is easier to obtain agreement when the suppliers know that the final decision has not been made. Now is the time to be low context, regardless of the country in which you are working. Don't assume understanding, and be especially cautious about quality-related terms.

These discussions will also be helpful in that they will help to build a closer relationship. The better the supplier and the customer know each other, the better they will treat each other if there is difficulty. Some of the worst problems occur when there is a total arms-length relationship.

Here is a checklist of items to be sure are clear.

### Specifications

Even though you should have made sure that your specifications are clear to a person in another country, raise the issue again. Are there ASTM plating specifications? Are MIL-STD specifications mentioned? Are your own in-house workmanship specifications mentioned? Be sure that all of these additional specifications are at the supplier and, above all, that the supplier understands them.

## Quality Plan

Unless you plan to use air freight forever, do not plan to control supplier quality by incoming inspection at your receiving site. Work with the remaining suppliers to understand how they propose to control their production processes. Know how and where they will measure and test your product to be sure that their manufacturing processes are under control.

## Marking

Does the supplier understand how to mark the country of origin on the part or its package? How permanent will the label be?

## Packaging

How do you want the supplier to package the part? If you are going to leave the important issue of package design to the supplier, you should still understand the general approach that the supplier will take.

## Warranty

Does the supplier understand the terms of your warranty requirements? If, for example, you take possession Ex works, will you still want the supplier to pay for return transportation from the United States?

## Flexibility and Forecasts

Does the supplier understand the use of your forecasts? How binding are they? If you want an upside capacity guarantee, how much over forecast do you expect the supplier to deliver? How much work in process will the supplier maintain? Will he or she be able to accommodate an engineering change without too much difficulty or expense? Will you be insisting on the right to "cover" (obtain alternative parts with your supplier covering incremental costs) if the supplier cannot ship, and does the supplier understand this?

## FINAL DECISION

Once you have been through the final discussions with the remaining suppliers, you should make your final decision. The major criteria will be the landed costs that you calculated and the ability that you and the supplier have to maintain a good buyer–seller relationship. You should, of course, tell the unsuccessful candidates that you have decided to purchase elsewhere. Since they were obviously close candidates, you do not want to leave hard feelings behind. You may be seeing them again.

This is also a good time to tell the successful candidate how you will be measuring his or her performance as a supplier, and what he or she can expect from you in terms of feedback and meetings.

## CONTRACT SIGNING

You should definitely have a written document describing the expectations of both parties, and both parties should sign it. The nature of this document will vary between countries. It may look like a contract when dealing with northern European countries. It may be a simple letter with generalities in some other countries, where a good relationship will be assumed. I suggest that you adjust your expectations in this matter. Do not insist on a rigid, bulletproof contract except in rare circumstances, such as forming a research partnership.

If you have the right supplier, you don't need a perfect contract. If you have the wrong supplier, it will be of very little use.

## *KEY POINTS*

- ▶ Potential suppliers who are likely to produce poor quality, be inflexible, or ship late should be eliminated before considering costs.
- ▶ Before choosing a supplier, be sure to develop a total "landed cost." This includes not only the cost of the part, but inventory cost, duties, customs clearance, and freight.
- ▶ After developing a landed cost for each of a small number of potential suppliers, carefully check some risks. Four key

risks are exchange rate change, unexpected air freight, scrap or rework due to design changes, and extra procurement overhead to manage a supplier. Political instability is an unquantifiable and sometimes overstated risk.

▶ Sign a contract or at least a document that describes the expectations of both parties.

*Chapter Twenty-Four*

# Managing Remote Suppliers

You are now in a business arrangement with a remote supplier. Unless you are working through the supplier's subsidiary or representative, all your contacts with the supplier will be long-distance. You now have the challenge of managing the relationship with foreign suppliers.

## KEY SUPPLIER-MANAGEMENT SKILLS

Because there are only a few differences between domestic and foreign supplier management, I'll list what I regard as good supplier-management practices that can be used for any supplier.

### *Relationship Management*

You and your organization should know how to classify supplier relationships into either routine buyer–seller relationships or long-term strategic relationships, and to handle both effectively. Long-term strategic relationships are becoming more important as technical complexity increases and as companies buy at higher and higher levels of integration. A company committing to a microprocessor family or to an application-specific integrated circuit (ASIC) design today is putting a good deal of trust in a supplier, and the supplier is committing a share of its resources to the customer. This is a long-term, strategic relationship.

Raw material suppliers can also be strategic if they are in a monopoly or near-monopoly position. Maintaining strategic relationships requires communication, mutual commitment, and mutual willingness to pass up occasional short-term advantage.

However, not all relationships are long-term or strategic. Commodity products with multiple, easily changeable suppliers lead to another kind of relationship. It need not be any less ethical or personalized, but it is a shorter-term relationship that leads to different supplier-management techniques.

Skilled supply managers need to know how to differentiate between the two types and how to manage both. A lot of misunderstanding and resentment can be developed by making policy statements implying that all supplier relationships are partnerships and then treating some suppliers as commodity suppliers. Similar problems develop when a supplier claims a partnership relationship and does not follow through.

## Defined Core Purchasing Values

Your organization should have a set of core purchasing values, describing what you look for in your supplier relationships. These values go by different names in different companies. Hewlett-Packard used Technology, Quality, Responsiveness, Delivery, and Cost (and recently added Environmental Responsibility). This became the acronym TQRDC-E and is now recognized shorthand for "What HP wants in a supplier."

Most companies' programs will emphasize cost, quality, on-time delivery, flexibility, and responsiveness to their customer (or to their customer's customer). The values should be well documented, publicized, and disseminated throughout the organization so that every buyer or supply manager uses the same terms and language.

## Supplier Measurement

Your organization should have a supplier measurement system in place that matches its core values. On-time delivery is a core value and easy to measure, once you define what "on time" means. Generally, it means that deliveries may be a few days early but must not arrive after the due date. You also need to establish where the delivery point is. Is it at your dock, or is it at the carrier's pickup point?

Quality is vital but is becoming increasingly difficult to measure as the quality of goods has improved dramatically over the last few years. Japanese companies have shown that they can achieve high quality at reasonable cost. Most American companies have learned this lesson also.

If your company can assemble a few hundred uninspected purchased parts into a manufactured product, turn it on or test it, and have it work the first time 98 to 100 percent of the time, you probably would not test or inspect components merely to rate suppliers. However, if you are having poor first-test results, you should be inspecting some goods at incoming inspection. You can use this data to measure the quality of each supplier.

Cost can be difficult to measure effectively. It needs to be measured against time, against other companies' prices, and against a worldwide base of potential suppliers. Your price intelligence search has to go beyond your country's borders or you will be getting incomplete data.

Other measurements, such as flexibility and responsiveness, tend to be more subjective but are still valuable. Ratings of flexibility and responsiveness give important information. You need to rate your suppliers against each other on these characteristics. This measurement has to be consistent or the data will not be meaningful.

### Regular Supplier Feedback

You should feed the results of the measurement back to the suppliers on a regular basis. You should show the suppliers where they stand on an absolute basis and where they stand against their competition. These rankings must have a visible effect on a supplier's business relationship. A commodity supplier should have an increase or decrease in business volume based on the rating. A long-term strategic supplier should be expected to identify and correct flaws when it has performance difficulties.

### Supply-Management Teams

Major buying decisions should be made by cross-functional teams whose members work well together and respect one another's contributions. In high-tech industries, buyers and engineers should work together in a team when they choose and manage suppliers. Often the buyer and engineer report to the same manager.

Design engineers should also be part of (and sometimes leaders of) the team, especially during new product development. At a minimum, purchasing should be active in new product development activities.

### Capacity Management

You should be able to develop purchasing forecast information and share it with your suppliers. Ideally, this is part-specific information that comes from material requirement planning (MRP) systems. This information should be shared immediately with your suppliers every time your master schedules change. If you do this, you should be able to keep lead times short and maintain adequate supply in a seller's market.

The real test of this system comes when there is a major increase in demand, either by your company or by your supplier's other customers. If you can obtain adequate supply, and lead times stay constant, you have a good system.

### Supply-Management Strategies

Major purchased items should have a documented purchasing strategy that is agreed upon, updated on an appropriate schedule, and followed. It will cover relatively stable information, such as the nature of the supplier's industry, the suppliers' key cost drivers, pricing strategies, and the ease of changing suppliers.

Other information changes more rapidly. Supplier performance can get better or worse; purchasing volumes increase and decrease. Entire countries become more or less competitive based on exchange rates. Your company should use the most recent information to project who the supply base should be and to formulate key goals for supply improvement.

Worldwide price information is a key part of the strategy. It should include supply and price information from countries around the world. The supply managers should know where in the world the low-cost, high-quality suppliers are today and where they are likely to be a few years in the future.

## GLOBAL SUPPLIER MANAGEMENT

Managing foreign suppliers requires the same tools as managing domestic suppliers. You need to have the fundamentals of a supplier measurement, ranking, and feedback system in place to manage either supplier. If you don't have the fundamentals, you are going to be making a lot of sudden, tactical moves to head off crisis situations with either kind of supplier.

Sudden moves with foreign suppliers cause more problems because both parties often need a longer planning horizon. You might be using ocean freight, which adds a few weeks to the delivery cycle. You might be working with a supplier whose culture has less tolerance for uncertainty. In either case, you will have to plan and act farther ahead than you would with a domestic supplier.

## EVALUATE SENSIBLY

Measure and evaluate suppliers on only the items that they are responsible for. If you have responsibility for the freight movements, do not hold the supplier responsible for on-time delivery to your dock. Measure instead on-time delivery to the carrier. If you are having difficulty with the freight carrier to the point where the supplier's business is in jeopardy, tell the supplier about the problem. A good supplier can probably offer suggestions.

## MAINTAIN PRESSURE ON COSTS

Measure price changes and competitiveness in both the supplier's currency and in U.S. dollars. Your supplier needs to know that it is in a worldwide competition for your business. The supplier may not feel responsible for currency movements, but needs to hear the unwelcome news that it might have to relocate manufacturing or change its own suppliers.

In addition, you should measure the supplier's year-to-year price and cost change in its own currency. The supplier should feel responsible for keeping the price dropping, or at least increasing more slowly than its competitors. If the supplier is in a large country, there may be industry-specific price indexes. These will show year-to-year industry performance in reducing price or controlling price inflation. You should expect your suppliers to do better than an industry average.

## INFORMAL COMMUNICATIONS

Your supplier should be getting a steady stream of relatively informal communications from you (and your IPO if you have one). You should be sending purchasing forecasts so that the supplier can plan for your volume. If you have performance data on quality and delivery, that should get to the supplier immediately. Communication should be two-way. The supplier may want to inquire about your business and might want to offer suggestions that will reduce costs for both of you. You need a steady communication and a solid personal relationship to handle this. Most of the arrangements that go badly astray do so because of lack of personal relationships between the parties.

Here are some of the issues that should be covered in informal communications.

## *Holiday Schedules*

Holidays cause problems when they are unexpected. You and your supplier should have a holiday schedule for each other's country. Ask the supplier for one and provide one to him or her.

## *Telephone*

You should know the phone numbers of key people at the supplier, and these people should know yours. Your use of the telephone will vary with the language skills you and your contacts at the supplier have. If both of you have trouble speaking the other's language, you probably will not be talking much on the phone.

Because of time differences, you probably will be making and receiving some phone calls from home. You should give the key people at your suppliers your home phone number. You should get their home phone numbers also. However, the difficulty in doing this will vary by the degree of privacy common in a country's culture. Keep in mind that in some countries it is very difficult for an employee to identify individual long-distance calls to get reimbursed by an employer. Japan and Germany are two examples. Phone subscribers there receive a lump-sum long-distance bill.

You will probably reach non-English-speaking family members occasionally. This is usually not pleasant for either party. If this happens often, you should try to learn enough of the supplier's language to give your name, the name of the person you are calling, and a request to call back. Be sure to pass on a big "thank you" through your contact to those family members, who are truly making a large step to accommodate you.

Finally, be aware of daylight saving time changes. Other countries do not change on the same schedule that we do. The unexpected hour difference one way or the other can be awkward.

## *Fax*

Fax is excellent for sending pictures and other graphic material. It has the drawback, however, that faxed messages are difficult to distribute internally in a company that is not used to handling large volumes of paper. Most Asian countries are used to handling faxes. Their written languages do not work well on keyboards, so the custom is to write a message longhand and fax it.

## *E-mail*

If your company and the supplier both have an e-mail system, investigate ways to tie your systems together. Most e-mail systems can send and receive external messages. They have links to either the Internet system or to a public e-mail system such as AT&T's x.400 system.

## FORMAL COMMUNICATION

Besides the informal contacts, you should be sure that there is ongoing, more formal contact between your company and the supplier. This should become a regular event. This is particularly important in low-context countries, or high-uncertainty avoidance countries. The formality and regularity of the meetings will enhance communications in those places.

### *Meetings*

Set up a regular schedule of review sessions, either quarterly or semiannually. It's a good idea to alternate locations between your site and the supplier's site. Even if you are buying from a domestic representative or subsidiary, you still need these meetings with the supplier's staff so that they can receive performance feedback directly.

In these review sessions, you will present the supplier with measurements of its performance against its key competitors. This ranking will include delivery, price, service, and, if possible, quality. In the United States, this is usually an opportunity for the supplier to make comments about the buyer's performance also. A supplier can often make suggestions that will reduce costs. In Asian societies, this kind of feedback will be more difficult to obtain, as a seller may be reluctant to criticize a buyer. It may take some time to draw out this valuable information.

## MAINTAIN CONTINUITY

Work to make your company's personnel changes as smooth as possible. If you are moving to a new responsibility, try to introduce your successor personally. If you can't do this, at least have a joint telephone conversation. Remember that U.S. job mobility causes concern in many other countries.

## ETHICS

Maintain a high level of ethical standards in dealing with foreign suppliers. You not only represent your company, you represent to some extent your whole country. Word of unethical practices will spread quickly in a small country, and life will become more difficult for you and others who are trying to do business there.

One of the practices that I find troublesome is that of cutting back on remote suppliers before cutting back on local suppliers. The suppliers in foreign countries are no less worthy, and often more needy, of your business than U.S. suppliers. If you must reduce a supplier base, do it based on performance, not distance.

With these few simple changes, management of remote suppliers can be as easy as managing local suppliers.

## *KEY POINTS*

▶ Sound domestic supplier management skills are necessary for foreign-supplier management also. Additional skills are required so that a buyer can minimize problems that can result from distance, from language, and from cultural misunderstandings.

▶ Proper management of a remote supplier requires an effective flow of both formal and informal communications. Both are necessary if an effective relationship is to be maintained.

*Chapter Twenty-Five*

# Conclusion

## SUMMARY

This book has covered the important differences between management of a global supply base and a domestic supply base. The keys to success with global suppliers are:

- Know how to "shop the world" for the world's best suppliers. Do this periodically as part of a formal product-procurement strategy.
- Do not let differences in culture, law, or currency stand in your way to success. Respect differences, but do not fear them.
- Do not treat distance as a significant factor, except as it affects delivery transit time.
- If your finance staff is not comfortable working with foreign currencies, lead them to a comfort level.
- Develop a working vocabulary and the ability to work with experts in logistics and customs.
- Take charge of your order and communication channels. Reduce dependence on intermediaries that do not work for you.
- Use the same supplier-management techniques on remote suppliers that you use on local suppliers, with some minor modifications.

## GETTING STARTED

While I hope this book has taken some of the mystery out of international procurement, no book is a substitute for experience. The best way to get started is to choose one of your two or three most strategic purchased products. If you start with a small, nonstrategic product, you will likely have difficulty obtaining sufficient company resources to be successful. If, somehow, you are successful, no one may notice.

If there is no product procurement strategy for the strategic product, start to write one. If you have a strategy, but it is weak on analyzing the supply base worldwide, correct this.

If there is a strategy and a worldwide analysis already, but you are dealing with remote suppliers through their U.S. subsidiaries and representatives, start by taking better control of the order and communication channels. Start to develop a relationship with the key people at the suppliers' headquarters. Consider direct ordering, and establish the capability of your logistics and customs departments.

## AND FINALLY,

Dealing with remote suppliers adds an extra dimension to the purchasing role. It leads to more-interesting work, wider horizons, and a greater insight into how the world works. Successful global supply managers will approach these tasks with an open mind and an adventurous spirit. The world is your market. Get there before your competitors do.

## Appendix A

# Answers to Tests of Understanding

## CHAPTER 9

1. Today's dollar equivalent is $100,000. Foreign currency divided by exchange rate equals dollars, DM170,000/1.7 = $100,000.

2. The exchange rate will be 1.785. A higher exchange rate per dollar indicates a stronger dollar. 1.70 times 1.05 (105%) equals 1.785.

3. The exchange rate will be 1.615. A lower rate per dollar indicates a weaker dollar. 1.70 times 0.95 (95%) equals 1.615.

4. The supplier takes the exchange-rate risk by setting the price in dollars. You will pay $100,000 no matter what the exchange rate is. If the dollar is 5 percent weaker, the rate will be DM1.615 per dollar. Foreign currency equals dollars times the exchange rate; $100,000 times 1.615 equals DM161,500. He can exchange the $100,000 for DM161,500.

5. You are still paying $100,000 because the price is set in dollars. The new exchange rate is 1.785. The supplier can exchange the $100,0000 for DM178,500. (Foreign currency equals dollars times exchange rate.)

6. You take the exchange risk by agreeing to pay the supplier DM170,000 regardless of what happens to exchange rates. If the dollar weakens to 1.615 marks per dollar, you must pay $105,263 for the DM170,000. (Dollars equal foreign currency divided by the exchange rate; 170,000/1.615 equals $105,263.)

7. You are still paying the supplier DM170,000. At an exchange rate of DM1.785 per dollar, DM170,000 will cost you $95,238. (Dollars equal foreign currency divided by the exchange rate; 170,000/1.785 equals $95,238.)

## CHAPTER 11

1. Dollars equal foreign currency divided by the exchange rate: $NT300,000/25.828 equals $11,615.

2. The first of the two rates is the rate for selling the foreign currency: $NT300,000/25.758 is $11,647. Banks will charge you the higher amount of dollars and give you the lower amount of dollars.

3. Dollars equal foreign currency divided by the exchange rate: 100,000/6.240 equals $16,026.

4. Lowering the interest rates in a country generally weakens the currency. You should be able to get more marks for your dollars.

5. A country selling its currency for another is trying to weaken its currency. Heavy selling of anything reduces its value.

## CHAPTER 12

1. German interest rates were lower than U.S. rates. The dollar was weaker against the German mark on the forward market than it was at spot. This indicates Germany had lower interest rates.

2. Dollar price equals foreign price divided by exchange rate. Use the 90-day forward rate of 0.6284: £600,000/0.6284 equals $954,806.

3. A buyer needs the foreign currency to pay suppliers, so the buyer contracts to buy the foreign currency.

4. Your banker will charge you £600,000/0.6244, or $960,922.

## CHAPTER 13

1. You exercise the option because you can get more deutsche marks per dollar through the option at 1.7 than you can at the spot rate of 1.60. You pay 500,000/1.7 for the DM (dollars equal foreign currency divided by exchange rate). This is

$294,118. In addition, there is a $35,000 option premium, giving a total cost of $329,118.

2. The spot rate on the expiration date is the same as the option strike price. There is no difference in cost between exercising the option and buying on the spot market. Therefore, the total cost is the same as in question 1, $329,118.

3. You do not exercise the option because you can get more deutsche marks per dollar on the spot market at 1.80 than you can with the option at 1.70. The deutsche marks cost 500,000/1.80, or $277,778. The option premium is $35,000, for a total cost of $312,778.

## CHAPTER 14

1. The contract costs $1,411,765 ($V_F$/C: ¥120M/85). It is worth, on the maturity date, $1,263,158 ($V_F$/S: ¥120M/95). You lose the difference between what it cost and what it's worth, $1,411,765 − $1,263,158, or $148,607 on the contract.

2. You bought 30 percent more product than you hedged, and the dollar strengthened by 15 percent. You will have paid approximately 4.5 percent less (see Table 14–1) than if you had forecasted accurately.

3. Because the dollar strengthened, you would not exercise your option. Therefore, you buy the entire volume at spot, exactly as you would have if you forecasted accurately. You paid 15 percent less for the parts than you expected, not including the cost of the option premium.

4. You buy 70,000 units at ¥120 apiece, so you pay the supplier ¥8,400,000. The spot rate that day is ¥95 per dollar, so the ¥8.4M costs you $88,421 (8,400,000/95). You then must purchase ¥12,000,000 at the rate of ¥84 per dollar to honor your forward contract. This costs you $142,857 (¥12M/84). The ¥12M you just bought is worth $126,316 (¥12M/95). You had a loss on the hedge transaction of $16,541 ($142,857 minus $126,316). Your total cost is $88,421 plus $16,541, or $104,962. Your unit cost is $104,962 divided by 70,000 units, or $1.496.

## CHAPTER 15

If the dollar strengthens 12 percent from the base exchange rate ($R_B$) of £0.65 per dollar, it will be worth 0.65 times 1.12, or £0.728. This gives the new exchange rate ($R_N$). The base price is £1.20. Using the formula

$$\text{New Price in Manufacturer's Currency} = \frac{\text{Base Price} \times 2 \times R_N}{R_B + R_N}$$

the new price is £1.2679. Your new dollar cost is $1.742, calculated by dollars equal foreign currency divided by the new exchange rate: £1.2679/ 0.728 = $1.742.

# Supplier Survey
*From the text of Global
Supply Management*

This survey is designed to be filled out by suppliers and buyers jointly. It is not intended to be mailed to a supplier. It is intended as one of the last steps in supplier selection, not one of the first. Total time to discuss and fill out is 8–12 hours. The material in this appendix may be copied from the book.

## PART ONE: GENERAL INFORMATION

Name of company: _____

Name of factory: _____

Date of survey: _____

Scope of survey: Whole plant _____ Product _____

1.   Address of factory: _____

City: _____

State or equivalent: _____

Country: _____

Telephone number: _____

English spoken? _____

Fax number: _____

English read/written? _____

2.   Headquarters address:   _____

                    City:   _____

      State or equivalent:   _____

                 Country:   _____

3.     Form of ownership:   _____

      Name of controlling
      owner, if applicable:   _____

            Year founded:   _____

## *Management Names and Years in Position*

President/Managing Dir.:   _____   Years_____

         R&D Manager:   _____   Years_____

Manufacturing Manager:   _____   Years_____

      Quality Manager:   _____   Years_____

      Finance Manager:   _____   Years_____

        Sales Manager:   _____   Years_____

## *Factory Information*

| Plant Location | This Plant | Other _____ | Other _____ |
|---|---|---|---|
| Size (in square meters) | | | |
| Total number of employees | | | |
| Administration employees | | | |
| Engineering employees | | | |
| Manufacturing employees | | | |
| Quality employees | | | |

Are employees
union members? _____

Which union? _____

When was last
work stoppage? _____

How long did it last? _____

## *Financial Data*

Currency _____        Fiscal year ends _____

| | Current Year | Previous Year | Year Before |
|---|---|---|---|
| Sales | | | |
| Profit after tax | | | |
| Assets | | | |
| Receivables | | | |
| Cash/equivalent | | | |
| Inventory | | | |
| Plant & equipment | | | |
| Other assets | | | |
| Short-term debt | | | |
| Long-term debt | | | |
| Other liabilities | | | |
| Equity | | | |
| Interest payments | | | |
| | | | |

What is the projected sales each year for the next three years?

_____

_____

_____


Who are the five largest customers and what share of the company's sales do they account for?

Name _____   Percent _____

Name _____   Percent _____

Name _____   Percent _____

Name _____   Percent _____

Name _____   Percent _____


What support life can the company commit to if we purchase this company's products?

_____

_____

_____

_____

_____


Does the company have major multinational customers? Who are they? Has it won performance awards from those customers?

_____

_____

_____

_____

_____


Answers for this section have been provided by:

Name: _____   Date: _____

## PART 2: MANUFACTURING

Company name _____ Plant _____

This report covers: Entire plant _____ or Product _____

1. What is the value of work-in-process inventory, in weeks of production?

   _____

   _____

   _____

2. How quickly can the status of a job be determined starting with a purchase order number?

   _____

   _____

   _____

3. What drawings accompany a job during production? Who makes the drawings?

   _____

   _____

   _____

4. Is there a defined process for implementing design changes occurring during manufacture?

   _____

   _____

5. Is there a separate place for customer-owned tooling?

   _____

   _____

6. Is the manufacturing area covered by fire-extinguishing sprinklers?

   _____

   _____

7. Is there a separate area for defective parts or tools?

_____

_____

8. Which of the following are measured? What are the most recent results, the previous year's results, and the goal for next year?

|  | One Year Ago | Current Level | Next-Year Goal |
|---|---|---|---|
| Scrap and rework costs as a percent of manufacturing costs |  |  |  |
| On-time shipment to customers |  |  |  |
| Cost reductions, percent of manufacturing cost |  |  |  |

Definition of "on time": _____

9. What processes are subcontracted? To whom?

_____

_____

_____

_____

10. How are work instructions provided to manufacturing employees?

_____

_____

_____

_____

11. Are all manufacturing and test procedures done under proper electrostatic discharge control conditions? What techniques are used? Did you observe any noncompliance by operators, supervisors, or visitors?

_____

_____

_____

_____

12. What is the age of this building?

_____

_____

13. Attach a list of all production equipment that will be used to produce this product. Include brand, model number, age, and whether computer-controlled. Identify stations where operators perform statistical process control.

_____

_____

_____

_____

_____

_____

_____

_____

_____

_____

_____

Answers for the manufacturing section were provided by:

Name: _____    Date: _____

Company name: _____    Plant: _____

## PART THREE: QUALITY

### *General Questions*

1. To whom does the senior in-plant quality manager report?

   _____

2. To whom does the companywide quality manager report?

   _____

3. Has this plant been qualified to ISO9001, 9002, or 9003?

   _____

4. Is there a "quality manual"?

   _____

5. Is it available to customers?

   _____

6. Does the quality department approve new designs and documentation?

   _____

7. Is the quality department involved in operator training?

   _____

8. Is there a calibration program for measuring equipment? Is it consistently followed?

   _____

9. What equipment is sent outside for calibration?

   _____

   _____

10. Who is the calibration lab? What standards does that lab use?

   _____

   _____

   _____

**The following questions pertain to purchased material and suppliers.**

1. Do purchase orders contain specifications of the products to be purchased?

   _____

2. How and when does the company state its quality and reliability expectations to suppliers?

   _____

   _____

   _____

   _____

3. What level of quality (ppm) is currently being obtained?

   _____

   _____

   _____

4. Who measures the quality?

   _____

   _____

   _____

5. When the company inspects incoming materials, what sample plan is used? (For example, LTPD, AQL, AOQL)

   _____

   _____

   _____

6. What is the procedure when discrepant material is detected at receipt?

   _____

   _____

   _____

   _____

   _____

   _____

## The following questions pertain to work in process.

1. At what points of building our product will in-process inspections be done?

_____

_____

_____

_____

2. Who performs this inspection? (Manufacturing or QA)

_____

_____

3. Is there a formal corrective action system to identify and correct in-process problems?

_____

4. What is the in-process failure rate overall?

_____

_____

_____

_____

_____

5. What was it a year ago?

_____

_____

_____

_____

_____

6. What is the goal for next year?

_____

_____

_____

_____

7. Are there general workmanship standards covering processes such as painting and soldering?

_____

| Process | Standard Name |
|---------|---------------|
|         |               |
|         |               |
|         |               |
|         |               |
|         |               |
|         |               |
|         |               |

8. What is the yield at first electrical or mechanical test done after product is completed?

_____

_____

_____

9. What was the yield one year ago?

_____

_____

_____

10. What is the goal for next year?

_____

_____

_____

_____

_____

_____

_____

## The following questions pertain to outgoing product quality.

1. What is the company's current measurement of its outgoing quality?

   _____

   _____

   _____

   _____

   _____

2. What was it a year ago?

   _____

   _____

   _____

   _____

   _____

3. What is the goal for next year?

   _____

   _____

   _____

   _____

   _____

4. Who makes the measurements: customer or this company?

   _____

5. Is the last test or inspection done on all products or on samples? If samples, what sampling plan is used?

   _____

   _____

   _____

   _____

6. Is the test done before or after final packaging?

_____

7. What is the customer return rate, expressed in units returned divided by units shipped?

_____

_____

_____

_____

_____

8. Does the company maintain reliability data on its products?

_____

9. What is the predicted reliability?

_____

_____

_____

_____

_____

10. What do customers report?

_____

_____

_____

_____

_____

_____

_____

The answers for this section have been provided by:

Name: _____ Date: _____

## PART FOUR: MATERIALS

1. Does purchasing respond to customer-generated forecasts, internal forecasts, or purchase orders?

   _____

   _____

2. What software is used for MRP?

   _____

   _____

   _____

3. Does the company share forecasts with its suppliers in order to control lead time?

   _____

4. What is the goal for inventory in weeks of supply?

   _____

   _____

   _____

   _____

5. Are supplier quality history and ranking used to select suppliers?

   _____

6. Are supplier on-time delivery and ranking used to select suppliers?

   _____

7. What raw materials are kept in climate-controlled areas?

   _____

   _____

   _____

   _____

8. Does the company have a formal change notice system that requires its suppliers to notify it of changes to purchased parts?

    _____

9. What imported materials will be used in our product? List type of component, supplier, and country of origin.

    _____

    _____

    _____

    _____

    _____

    _____

    _____

    _____

    _____

    _____

    _____

    _____

    _____

Answers for this section were provided by:

Name: _____    Date: _____

*Appendix C*

# Buyer's Guide
# to Key Countries

This appendix gives key information for 14 countries that either are, or are about to become, major exporters to the United States. It contains the following information.

## MAJOR EXPORTS TO THE UNITED STATES, 1994

These are each country's major exports to the United States and the growth rate between 1993 and 1994. Products are listed by four-digit HS Codes. The data is courtesy of Global Trade Information Services, Inc., of Columbia, South Carolina.

## TRADE REPRESENTATION IN THE UNITED STATES

This gives the telephone and fax numbers of the country's embassy and consulates or trade offices that will help buyers.

## USEFUL CONTACTS IN THE SUPPLIER'S COUNTRY

This gives the phone and fax numbers of the American Chamber of Commerce, some accounting firms, and a law firm. I have selected some large law and accounting firms that have many offices out of the United States. (This is not intended to be an endorsement of the abilities of any particular firm, however.)

The law firm I selected was Baker and McKenzie. If they do not have an office in a particular country, I listed either another American firm or a large local firm.

The accountants I selected were Arthur Andersen, KPMG Peat Marwick, and Price Waterhouse.

Unless indicated, the offices listed are in the country's capital city. In some cases offices in a major commercial city were listed instead.

## *Foreign Phone Numbers*

The numbers are listed as dialed from outside the listed country. The first character is a "+", which indicates that you should dial the international access code used in the country you are dialing from. (In the United States, the access code is 011.) The first group of figures is the country code, and the second is the city code. (Singapore is an exception; it has no city codes.) The remaining figures are the telephone number. If you enter a "#" sign after the telephone number when dialing from the United States, the call will go through faster. (The "#" sign signals the international dialing circuitry that no more numbers are coming.)

If you call from inside the country, you do not dial the international access code or the country code. If you are calling from inside the country but outside the city, insert a "0" (zero) as the first figure in the city code. (Canada and the United States are exceptions to this practice.)

## BUYER'S CULTURAL RADAR DIAGRAM

The radar diagram shows the six most important cultural characteristics that affect a buyer's relationship with a potential seller.

## *Power Distance*

Businesses and people in countries with a higher power distance score will:

- Have less employee participation in decision-making and quality programs.
- Be more likely to have one key decision maker.
- Make faster decisions, but often with slower implementation.
- Demonstrate levels of power and authority that may make buyers uncomfortable.

Businesses and people in countries with lower power distance scores will:

- Have diffuse decision making, often bottom-up.
- Generally implement decisions faster.
- Be uncomfortable with buyers' displays of power or authority.

### Individualism

The United States has an extremely high individualism score. Businesses and people in countries with significantly lower scores will:

- Have a higher degree of conformity in dress and behavior.
- Be motivated more by group (family or company) success than individual success.
- Be uncomfortable with one person being praised or blamed publicly.
- Be uncomfortable with too individualistic behavior by buyers.

### Uncertainty Avoidance

Businesses and people in countries with higher uncertainty avoidance scores will:

- Have a higher level of ritualized social behavior.
- Be more disturbed by surprise moves by buyers.
- Be less flexible in factory scheduling.
- Want more certainty in contracts or in human understanding of business partners.

Countries with lower uncertainty avoidance scores will:

- Be more tolerant of buyer-required schedule changes.
- Be less detail oriented in contracts.

### Context

Businesses and people in countries with higher context scores will:

- Require a higher degree of networking between individuals and departments in order to get to a decision.
- Be unused to receiving information in long, detailed briefings.

- Need to get to know buyers on a personal level before being comfortable doing business.
- Appear cryptic and incomplete in their messages to you.

Businesses and people in countries with lower context scores will:

- Give longer, more thorough messages to you.
- Be used to receiving information in formal briefings.
- Be less concerned with personal relationships and more concerned with business relationships.
- Not share information automatically between people and departments.


### *Buyer Rank*

Businesses and people in countries with a higher buyer-rank score will:

- Be more agreeable to meeting properly presented buyer needs.
- Show a great deal of deference toward buyers.
- Be more likely to entertain visitors.

Businesses and people in countries with a lower buyer-rank score will:

- Be less automatically agreeable to buyer needs.
- Resent attempts by buyers to use economic power.
- Be less likely to entertain visitors.


### *Monochronism*

Businesses and people in countries with a higher monchronism score will:

- Be more prompt for meetings and expect buyers to do the same.
- Have private, uninterrupted meetings.
- End meetings on a schedule, regardless of completion of the business.
- Have longer lead times in factories.
- Be more likely to meet delivery commitments.

Businesses and people from countries with a lower monochronism score will:

- Be more likely to complete a meeting or other transaction and risk being late for the next meeting.
- Be more likely to be late and more tolerant of others' lateness.
- Be more likely to have interruptions during meetings.
- Have more-flexible factory scheduling and shorter lead times.
- Be more likely to miss promised shipment dates.

## CURRENCY HISTORY

In order to judge currency stability, I graphed the exchange rate of the country's currency for a number of years. I graphed the values at the start of each month, so total fluctuations will be slightly larger.

## PURCHASER'S GUIDE TO CANADA

*Major Exports to the United States, 1994*

| HS Code | Description | Value, Million Dollars | Growth Rate 94/93 |
|---|---|---|---|
| 8703 | Cars | $22,710 | 23.6% |
| 8704 | Trucks | 6,699 | −11.2 |
| 8708 | Vehicle parts and accessories | 6,453 | +7.9 |
| 4407 | Lumber | 5,669 | +19.2 |
| 2709 | Petroleum from crude | 4,917 | −1.6 |
| 2711 | Natural gas | 4,878 | +16.6 |
| 4801 | Newsprint | 3,296 | −7.1 |
| 8473 | Office mach. parts and accessories | 2,105 | +53.2 |
| 7601 | Unwrought aluminum | 2,053 | +38.3 |
| 4703 | Wood pulp | 1,693 | +24.4 |
| 2710 | Petroleum not from crude | 1,577 | −6.0 |
| 7108 | Gold | 1,489 | −5.8 |
| 8701 | Tractors | 1,402 | +114.4 |
| 8542 | Integrated circuits | 1,313 | −2.2 |
| 8407 | Spark-fired int. combustion engines | 1,124 | −6.3 |
| 8480 | Molds for metalworking | 1,120 | +198.6 |
| 4802 | Paper | 1,117 | +9.2 |
| 8471 | Computers and computer parts | 1,069 | +15.5 |

Source: Global Trade Information Services, Inc., Columbia, South Carolina.

## Trade Representation in the United States

| Organization | Phone | Fax |
|---|---|---|
| Embassy, Washington | (202) 682 1740 | (202) 682 7726 |
| Consulate, Chicago | (312) 616 1860 | (312) 616 1877 |
| Consulate, Dallas | (214) 922 9806 | (214) 922 9815 |
| Consulate, Detroit | (313) 567 2340 | (313) 567 2164 |
| Consulate, Los Angeles | (213) 345 2700 | (213) 620 8827 |
| Consulate, Miami | (305) 589 1600 | (305) 374 6774 |
| Consulate, Minneapolis | (612) 333 4641 | (612) 332 4061 |
| Consulate, Seattle | (206) 443 1777 | (206) 443 1728 |

## Useful Contacts

| Organization | Phone | Fax |
|---|---|---|
| Baker and McKenzie, Toronto | (416) 863 1221 | (416) 863 6275 |

Note: Major U.S. accounting firms are in cities throughout Canada.

## Cultural Radar Diagram

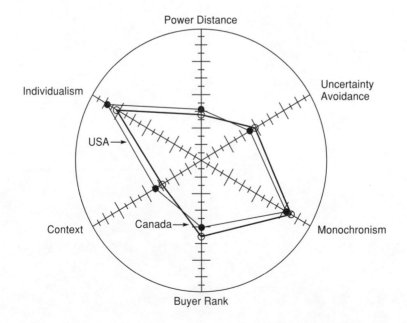

## *Currency History*

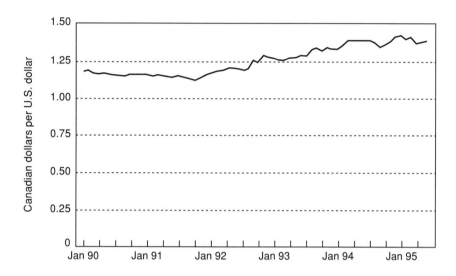

# PURCHASER'S GUIDE TO CHINA

## *Major Exports to the United States, 1994*

| HS Code | Description | Value, Million Dollars | Growth Rate 94/93 |
|---------|-------------|------------------------|-------------------|
| 9503 | Toys | $2,673 | +14.2% |
| 6403 | Leather shoes | 2,652 | +25.3 |
| 6402 | Rubber and plastic footwear | 1,764 | +3.0 |
| 4202 | Luggage and handbags | 1,550 | +18.2 |
| 8527 | Radio receivers | 1,391 | +48.9 |
| 6110 | Knitted sweaters, sweatshirts, vests | 1,171 | +4.8 |
| 6204 | Women's suits and ensembles | 931 | +3.8 |
| 4203 | Leather clothing | 916 | +49.0 |
| 9405 | Lamps and lights | 841 | +38.8 |
| 8525 | Transmitters, including phone | 828 | +99.2 |
| 8471 | Computers and parts of computers | 713 | +62.9 |
| 6206 | Women's nonknitted blouses | 689 | −24.7 |
| 3926 | Plastic nonhousehold items | 675 | +19.2 |
| 9502 | Humanlike dolls | 654 | +17.9 |
| 6404 | Shoes with textile uppers | 653 | +25.3 |
| 9505 | Festive and entertainment products | 626 | +19.6 |
| 8473 | Office mach. parts and accessories | 591 | +89.5 |
| 9506 | Sporting goods | 568 | +65.6 |
| 8516 | Electric heaters | 558 | +13.6 |
| 9504 | Games | 556 | +60.3 |

Source: Global Trade Information Services, Inc., Columbia, South Carolina.

## *Trade Representation in the United States*

| Organization | Phone | Fax |
|--------------|-------|-----|
| Embassy, Washington | (202) 328 2500 | |
| U.S.-China Business Council | (202) 429 0340 | |

## *Useful Contacts*

| Organization | Phone | Fax |
|---|---|---|
| American Chamber of Commerce | +86 1 500 7766 | |
| Arthur Andersen | +86 1 505 3333 | +86 1 505 1828 |
| Baker and McKenzie | +86 1 505 0591 | +86 1 505 2309 |
| KPMG Peat Marwick | +86 1 501 3388 | +86 1 500 4059 |
| Price Waterhouse | +86 1 505 1524 | +86 1 505 1026 |
| U.S. Embassy | +86 1 532 3831 | |

## *Cultural Radar Diagram*

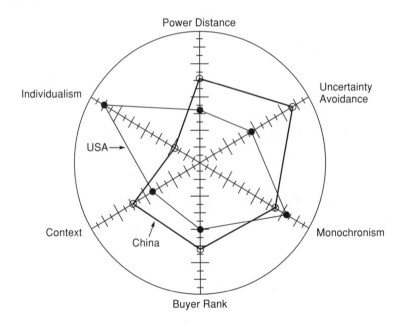

## *Currency History*

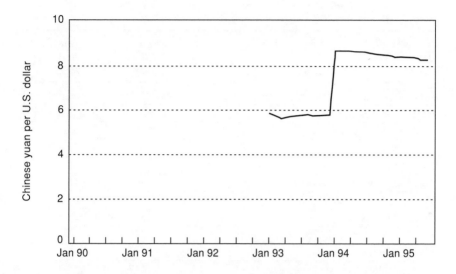

# PURCHASER'S GUIDE TO FRANCE

## Major Exports to the United States, 1994

| HS Code | Description | Value, Million Dollars | Growth Rate 94/93 |
|---|---|---|---|
| 8411 | Gas turbines and parts | $1,958 | −6.5% |
| 8802 | Aircraft and parts | 1,843 | −2.6 |
| 8708 | Parts and accessories for vehicles | 764 | +17.0 |
| 2204 | Wine | 511 | +2.4 |
| 3303 | Perfume and toilet water | 404 | −4.8 |
| 9701 | Paintings and drawings | 401 | +18.7 |
| 8542 | Integrated circuits | 364 | +63.7 |
| 2208 | Liquors | 341 | +28.1 |
| 8803 | Parts of aircraft | 290 | +22.3 |
| 2710 | Petroleum not from crude | 277 | +48.8 |
| 2844 | Radioactive elements | 177 | +55.8 |
| 8429 | Earth-moving equipment | 175 | +46.6 |
| 8471 | Computers and parts of computers | 167 | +5.4 |
| 9022 | X-ray equipment | 151 | −2.5 |
| 8473 | Office equipment parts and accessories | 139 | +19.6 |
| 7208 | Hot-rolled steel | 135 | +201.8 |
| 3304 | Beauty products | 122 | +0.8 |
| 8477 | Rubber-working or plastic-working machinery | 121 | +63.3 |

Source: Global Trade Information Services, Inc., Columbia, South Carolina.

## *Trade Representation in the United States*

| Organization | Phone | Fax |
|---|---|---|
| Embassy, Washington | (202) 944 6343 | (202) 944 6336 |
| Trade Office, Atlanta | (404) 522 4843 | (404) 522 3039 |
| Trade Office, Boston | (617) 523 4456 | (617) 523 4461 |
| Trade Office, Chicago | (312) 661 1880 | (312) 661 0976 |
| Trade Office, Detroit | (313) 567 0510 | (313) 567 3149 |
| Trade Office, Houston | (713) 266 7595 | (713) 266 3424 |
| Trade Office, Los Angeles | (310) 843 1700 | (310) 843 1701 |
| Trade Office, Miami | (305) 579 4783 | (305) 372 1863 |
| Trade Office, New York | (212) 397 8800 | (212) 315 1017 |
| Trade Office, San Francisco | (415) 781 0986 | (415) 781 4750 |

## *Useful Contacts*

| Organization | Phone | Fax |
|---|---|---|
| American Chamber of Commerce | +33 1 47 23 70 28 | +33 1 47 20 18 62 |
| Arthur Andersen | +33 1 42 91 06 06 | +33 1 42 91 09 90 |
| Baker and McKenzie | +33 1 44 17 53 00 | +33 1 44 17 45 75 |
| KPMG Peat Marwick | +33 1 46 39 46 39 | +33 1 47 58 81 21 |
| Price Waterhouse | +33 1 41 26 40 00 | +33 1 41 26 41 26 |
| U.S. Embassy | +33 1 42 66 48 27 | |

## *Cultural Radar Diagram*

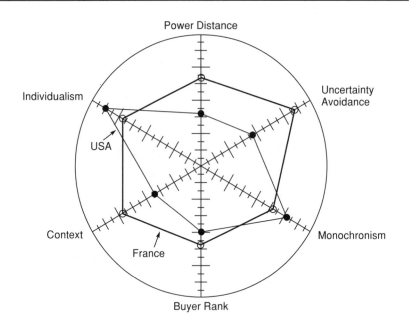

## *Currency History*

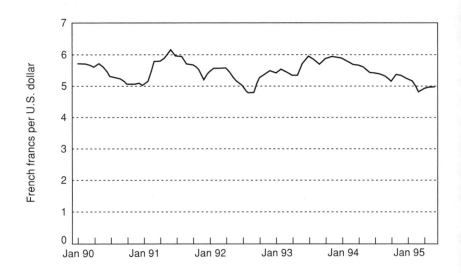

## PURCHASER'S GUIDE TO GERMANY

### Major Exports to the United States, 1994

| HS Code | Description | Value, Million Dollars | Growth Rate 94/93 |
|---------|-------------|------------------------|-------------------|
| 8703 | Cars | $5,800 | +6.8% |
| 8407 | Spark-fired int. combustion engines | 744 | +10.9 |
| 8708 | Parts and accessories for vehicles | 717 | +9.3 |
| 9018 | Medical instruments and parts | 477 | −9.0 |
| 8411 | Gas turbines and parts | 458 | +37.1 |
| 8443 | Printing machinery and parts | 456 | +10.0 |
| 8413 | Liquid pumps | 449 | +18.3 |
| 2933 | Heterocyclic comp. w/ N-het. atoms | 394 | +20.1 |
| 9022 | X-ray equipment | 382 | −14.3 |
| 8471 | Computers and parts of computers | 382 | +4.5 |
| 8483 | Transmissions and shafts | 378 | +21.6 |
| 8542 | Integrated circuits | 368 | +68.1 |
| 8479 | Miscellaneous machinery | 362 | +24.2 |
| 8536 | Switches, breakers, fuses | 361 | +30.6 |
| 8477 | Rubber-working or plastic-working machinery | 343 | +25.7 |
| 8481 | Valves | 328 | +18.8 |
| 8701 | Tractors | 299 | +27.1 |
| 8473 | Office machine parts and accessories | 297 | +46.1 |

Source: Global Trade Information Services, Inc., Columbia, South Carolina.

## Trade Representation in the United States

| Organization | Phone | Fax |
|---|---|---|
| Embassy, Washington | (202) 974 8830 | (202) 298 4249 |
| German-American Chamber of Commerce, Atlanta | (404) 239 9494 | (404) 264 1761 |
| German-American Chamber of Commerce, Chicago | (312) 664 2662 | (312) 664 0738 |
| German-American Chamber of Commerce, Houston | (713) 877 1114 | (713) 877 1602 |
| German-American Chamber of Commerce, Los Angeles | (310) 297 7979 | (310) 297 7966 |
| German-American Chamber of Commerce, New York | (212) 974 8830 | (212) 974 8867 |
| German-American Chamber of Commerce, Philadelphia | (215) 665 1585 | (215) 665 0375 |
| German-American Chamber of Commerce, San Francisco | (415) 392 2262 | (415) 392 1314 |

## Useful Contacts

| Organization | Phone | Fax |
|---|---|---|
| Baker and McKenzie | +49 69 299 080 | +49 69 2990 8108 |
| KPMG Peat Marwick | +49 69 9587 0 | +49 69 9587 1050 |
| Price Waterhouse | +49 69 152 040 | +49 69 1520 4107 |
| Arthur Andersen | +49 69 75 710 | +49 69 74 8096 |

Note: Frankfurt offices listed. Others available throughout Germany.

## *Cultural Radar Diagram*

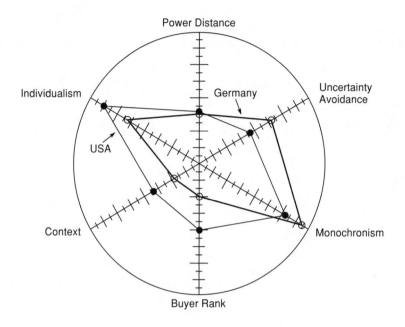

## *Currency History*

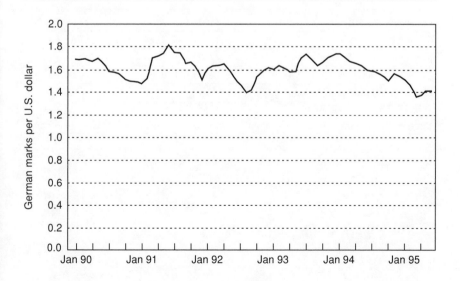

# PURCHASER'S GUIDE TO INDIA

## *Major Exports to the United States, 1994*

| HS Code | Description | Value, Million Dollars | Growth Rate 94/93 |
|---------|-------------|------------------------|-------------------|
| 7102 | Diamonds | $1,271 | +2.2% |
| 6204 | Women's suits and ensembles | 310 | +19.6 |
| 6206 | Women's nonknitted blouses | 297 | +9.6 |
| 7113 | Jewelry | 206 | +28.6 |
| 0801 | Coco-, cashew, and Brazil nuts | 175 | +30.2 |
| 6205 | Men's nonknitted shirts | 147 | +24.3 |
| 0306 | Crustaceans | 132 | +65.8 |
| 6211 | Athletic clothing | 130 | +40.8 |
| 4203 | Leather clothing | 111 | +7.2 |
| 5701 | Knotted carpets | 93 | +2.3 |

Source: Global Trade Information Services, Inc., Columbia, South Carolina.

## *Trade Representation in the United States*

| Organization | Phone | Fax |
|--------------|-------|-----|
| Embassy, Washington | (202) 939 7000 | (202) 939 7027 |
| Consulate, Chicago | (312) 781 6274 | (312) 781 6269 |
| Consulate, New York | (212) 879 7840 | (212) 861 3788 |
| Consulate, San Francisco | (415) 668 0967 | (415) 668 2073 |
| Electronics Trade and Technology Development Corporation | (714) 557 2073 | (714) 545 2723 |
| Engineering Export Promotion Council | (312) 236 2162 | (312) 236 4625 |

## Useful Contacts

| Organization | Phone | Fax |
|---|---|---|
| American Business Council | +91 11 688 5443 | +91 11 688 5046 |
| Arthur Andersen | +91 22 218 2929 | +91 22 218 0290 |
| Ashok C. Pratap and Co. (Attorneys) | +91 22 204 8090 | +91 22 204 2532 |
| KPMG Peat Marwick | +91 22 218 8010 | +91 22 218 8022 |
| Price Waterhouse | +91 22 283 5190 | +91 22 204 5592 |
| U.S. Embassy | +91 11 600 651 | |

Note: American Business Council and Embassy are in New Delhi. Other organizations are in Bombay.

## Cultural Radar Diagram

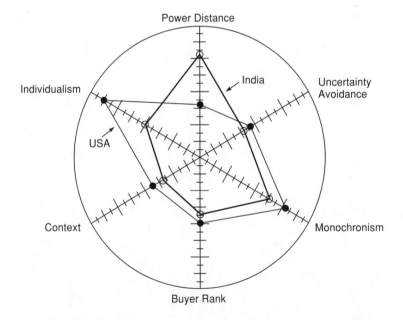

## *Currency History*

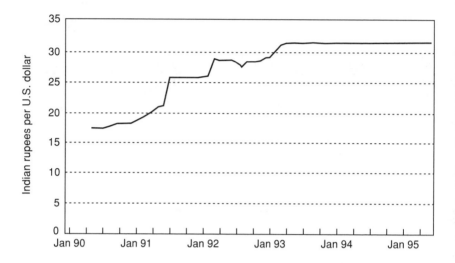

# PURCHASER'S GUIDE TO ITALY

## Major Exports to the United States, 1994

| HS Code | Description | Value, Million Dollars | Growth Rate 94/93 |
|---|---|---|---|
| 7113 | Jewelry | $1,306 | +1.8% |
| 6403 | Leather shoes | 739 | +17.3 |
| 8708 | Parts and accessories for vehicles | 382 | +30.5 |
| 9401 | Furniture, seats | 345 | +29.5 |
| 2204 | Wine | 281 | +16.3 |
| 8473 | Office mach. parts and accessories | 262 | +56.0 |
| 4202 | Luggage and handbags | 239 | +35.1 |
| 6203 | Men's suits | 236 | +21.8 |
| 9403 | Furniture | 209 | +31.8 |
| 9003 | Eyeglass frames | 206 | +10.3 |
| 2710 | Petroleum not from crude | 202 | −41.3 |
| 6908 | Glazed ceramic materials | 201 | +12.0 |
| 8803 | Parts of aircraft | 196 | −17.4 |
| 2941 | Antibiotics | 193 | −22.7 |
| 1509 | Olive oil | 188 | +8.7 |
| 6802 | Building stone | 183 | +10.4 |
| 8471 | Computers and parts of computers | 176 | +98.2 |
| 6204 | Women's suits and ensembles | 176 | +9.1 |
| 8481 | Valves | 169 | +32.9 |

Source: Global Trade Information Services, Inc., Columbia, South Carolina.

## Trade Representation in the United States

| Organization | Phone | Fax |
|---|---|---|
| Embassy, Washington | (202) 328 5500 | (202) 328 5538 |
| Italian Trade Commission, Atlanta | (404) 525 0660 | (404) 525 5112 |
| Italian Trade Commission, Chicago | (312) 670 4360 | (312) 670 5147 |
| Italian Trade Commission, Los Angeles | (213) 879 0950 | (310) 879 0950 |
| Italian Trade Commission, New York | (212) 980 1500 | (212) 758 1050 |

## *Useful Contacts*

| Organization | Phone | Fax |
|---|---|---|
| American Chamber of Commerce | +39 2 86 90 661 | +39 2 80 57 737 |
| Arthur Andersen | +39 2 290 371 | +39 2 657 2876 |
| Baker and McKenzie | +39 2 7601 3921 | +39 2 7600 8322 |
| KPMG Peat Marwick | +39 2 67 63 1 | +39 2 6763 2445 |
| Price Waterhouse | +39 2 778 51 | +39 2 778 5240 |

Note: All contacts are in Milan.

## *Cultural Radar Diagram*

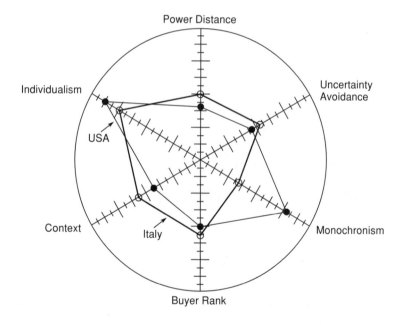

## *Currency History*

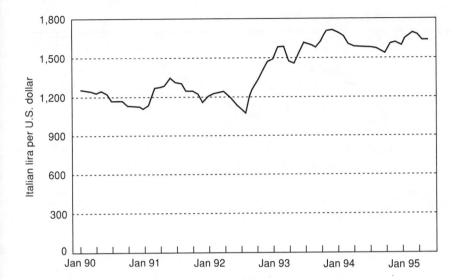

# PURCHASER'S GUIDE TO JAPAN

## Major Exports to the United States, 1994

| HS Code | Description | Value, Million Dollars | Growth Rate 94/93 |
|---------|-------------|------------------------|-------------------|
| 8703 | Cars | $24,367 | +11.0% |
| 8471 | Computers and parts of computers | 10,566 | +8.9 |
| 8708 | Parts and accessories for vehicles | 6,942 | +14.4 |
| 8542 | Integrated circuits | 6,841 | +30.8 |
| 8473 | Office mach. parts and accessories | 4,463 | +32.4 |
| 8407 | Spark-fired int. combustion engines | 3,069 | +26.3 |
| 8525 | Transmitters, including phone | 2,805 | +2.5 |
| 9009 | Copy machines and parts | 2,630 | +5.7 |
| 8704 | Trucks | 1,686 | +15.9 |
| 8517 | Tel. and fax equipment and parts | 1,606 | +4.0 |
| 9504 | Games | 1,372 | −46.3 |
| 8409 | Parts for spark or diesel engines | 1,214 | +21.5 |
| 8536 | Switches, breakers, fuses | 1,166 | +16.1 |
| 8521 | Video recorders | 1,166 | −9.7 |
| 8479 | Miscellaneous machinery | 1,146 | +25.0 |
| 8527 | Radio receivers | 984 | −7.7 |
| 8523 | Unrecorded media | 870 | +5.4 |
| 8429 | Earth-moving equipment | 862 | +37.7 |
| 8519 | Record, tape, video players | 836 | −10.2 |

Source: Global Trade Information Services, Inc., Columbia, South Carolina.

## *Trade Representation in the United States*

| Organization | Phone | Fax |
|---|---|---|
| Embassy, Washington | (202) 939 6700 | (202) 328 2187 |
| Consulate, Anchorage | (907) 279 8428 | (907) 279 9271 |
| Consulate, Atlanta | (404) 892 2700 | (404) 881 6321 |
| Consulate, Boston | (617) 973 9772 | (617) 542 1329 |
| Consulate, Chicago | (312) 280 0400 | (312) 280 9568 |
| Consulate, Detroit | (313) 567 0120 | (313) 567 0274 |
| Consulate, Honolulu | (808) 536 2226 | (808) 537 3276 |
| Consulate, Houston | (713) 652 2977 | (713) 651 7822 |
| Consulate, Kansas City | (816) 471 0111 | (816) 472 4248 |
| Consulate, Los Angeles | (213) 617 6700 | (213) 617 6727 |
| Consulate, Miami | (305) 530 9090 | (305) 530 0950 |
| Consulate, New York | (212) 371 8222 | (212) 319 6357 |
| Consulate, New Orleans | (504) 529 2101 | (504) 568 9847 |
| Consulate, Portland | (503) 221 1811 | (503) 224 8936 |
| Consulate, San Francisco | (415) 777 3533 | (415) 974 3660 |
| Consulate, Seattle | (206) 682 9107 | (206) 624 9097 |

## *Useful Contacts*

| Organization | Phone | Fax |
|---|---|---|
| American Chamber of Commerce | +81 3 3433 5381 | +81 3 3436 1446 |
| Arthur Andersen | +81 3 5228 4700 | +81 3 5228 4712 |
| Baker and McKenzie (affiliate) | +81 3 3403 5281 | +81 3 3470 3152 |
| KPMG Peat Marwick | +81 3 3578 1910 | +81 3 3434 2122 |
| Price Waterhouse | +81 3 3404 9351 | +81 3 3404 8610 |
| U.S. Embassy | +81 3 3224 5000 | |

## *Cultural Radar Diagram*

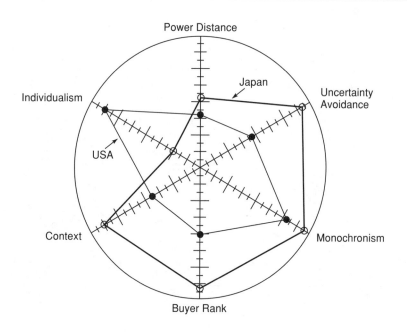

## *Currency History*

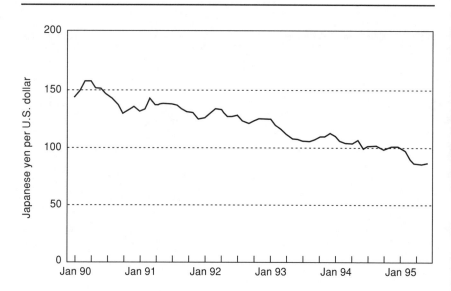

# PURCHASER'S GUIDE TO MALAYSIA

## Major Exports to the United States, 1994

| HS Code | Description | Value, Million Dollars | Growth Rate 94/93 |
|---------|-------------|------------------------|-------------------|
| 8542 | Integrated circuits | $3,114 | +27.0% |
| 8471 | Computers and parts of computers | 2,029 | +62.7 |
| 8527 | Receivers, including phone | 1,351 | +30.8 |
| 8521 | Video recorders | 732 | +16.5 |
| 8528 | Televisions | 439 | +48.3 |
| 8541 | Discrete semiconductors | 412 | +16.4 |
| 4015 | Rubber clothing | 402 | +13.5 |
| 8525 | Transmitters, incl. cordless phone | 398 | +33.7 |
| 8473 | Parts and accessories for computers | 372 | +76.4 |
| 8517 | Telephone and fax equip. and parts | 343 | +17.7 |
| 8519 | Record, tape players | 325 | +222.5 |
| 8520 | Tape recorders, inc. ans. mach. | 213 | +24.4 |
| 9006 | Cameras and parts | 176 | +36.6 |
| 9403 | Furniture, not seats | 154 | +59.5 |
| 9401 | Furniture, seats | 147 | +51.3 |
| 4001 | Natural rubber | 138 | +4.3 |
| 6205 | Men's nonknitted shirts | 134 | −5.9 |
| 8518 | Microphones, speakers, amps | 132 | +44.2 |
| 8531 | Sound and visual signaling equip. | 101 | +21.4 |

Source: Global Trade Information Services, Inc., Columbia, South Carolina.

## Trade Representation in the United States

| Organization | Phone | Fax |
|--------------|-------|-----|
| Embassy, Washington | (202) 328 2700 | (202) 332 8914 |
| Consulate, Chicago | (312) 787 4532 | (312) 787 4769 |
| Consulate, Los Angeles | (213) 621 2661 | (213) 620 8659 |
| Consulate, New York | (212) 682 0232 | (212) 983 1987 |

## *Useful Contacts*

| Organization | Phone | Fax |
|---|---|---|
| American Business Council | +60 3 248 2407 | +60 3 242 8540 |
| Arthur Andersen | +60 3 255 7000 | +60 3 255 5332 |
| David Chong & Co. (attorneys) | +60 3 263 2277 | +60 3 263 2278 |
| KPMG Peat Marwick | +60 3 255 3388 | +60 3 255 0971 |
| Price Waterhouse | +60 3 293 1077 | +60 3 293 0997 |
| U.S. Embassy | +60 3 248 9011 | +60 3 242 2207 |

## *Cultural Radar Diagram*

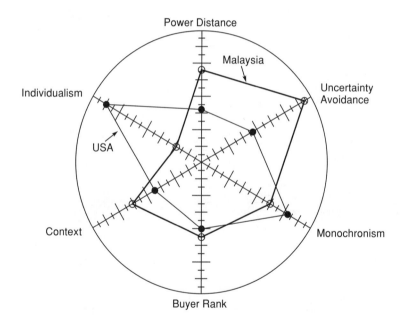

## *Currency History*

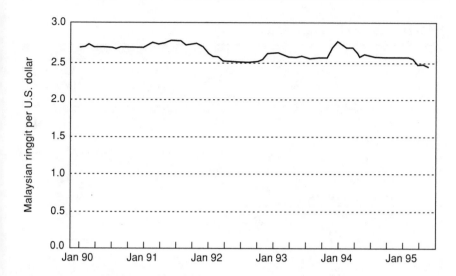

# PURCHASER'S GUIDE TO MEXICO

## Major Exports to the United States, 1994

| HS Code | Description | Value, Million Dollars | Growth Rate 94/93 |
|---------|-------------|------------------------|-------------------|
| 2709 | Petroleum from crude | $4,653 | +9.6% |
| 8703 | Cars | 3,944 | +27.9 |
| 8544 | Wire cable and assemblies | 2,969 | +30.0 |
| 8528 | Televisions | 2,265 | +42.5 |
| 8708 | Parts and accessories for vehicles | 2,180 | –3.0 |
| 8536 | Switches, breakers, fuses | 1,388 | +30.6 |
| 8407 | Spark-fired int. combustion engines | 1,331 | +64.1 |
| 8471 | Computers and parts of computers | 932 | +93.0 |
| 8527 | Receivers, including phone | 924 | +37.4 |
| 8529 | Parts for radio, TV, phones | 882 | +9.3 |
| 9401 | Furniture, seats | 805 | +31.5 |
| 8504 | Transformers | 705 | +3.1 |
| 8525 | Transmitters, incl. cordless phone | 653 | +165.0 |
| 8704 | Trucks | 643 | +18.4 |
| 8473 | Parts and accessories for computers | 599 | +20.4 |
| 8501 | Electric motors | 597 | +29.7 |
| 9029 | Speedometers and tachometers | 518 | +144.2 |
| 9032 | Automatic regulating equipment | 457 | +13.8 |
| 6203 | Men's suits and ensembles | 446 | +26.1 |

Source: Global Trade Information Services, Inc., Columbia, South Carolina.

## Trade Representation in the United States

| Organization | Phone | Fax |
|--------------|-------|-----|
| Embassy, Washington | (202) 728 1600 | (202) 775 4522 |
| Bancomext, Atlanta | (404) 522 5373 | (404) 523 2450 |
| Bancomext, Chicago | (312) 856 0316 | (312) 856 1834 |
| Bancomext, Dallas | (214) 688 4096 | (214) 905 3831 |
| Bancomext, Los Angeles | (213) 628 1220 | (213) 628 8466 |
| Bancomext, Miami | (305) 372 9929 | (305) 374 1238 |
| Bancomext, New York | (212) 826 2978 | (212) 826 2979 |
| Bancomext, San Antonio | (210) 525 9748 | (210) 525 8355 |

## Useful Contacts

| Organization | Phone | Fax |
|---|---|---|
| American Chamber of Commerce | +52 5 724 3800 | +52 5 703 3908 |
| Arthur Andersen | +52 5 326 8800 | +52 5 596 4692 |
| Baker and McKenzie | +52 5 230 2900 | +52 5 557 8829 |
| KPMG Peat Marwick | +52 5 726 4343 | +52 5 596 8060 |
| Price Waterhouse | +52 5 211 7883 | +52 5 286 6248 |

## Cultural Radar Diagram

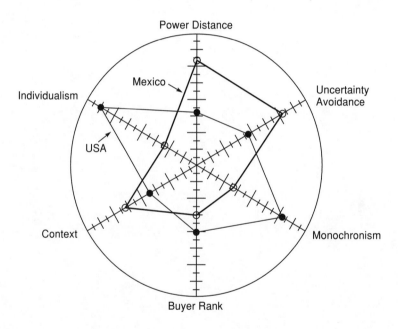

## *Currency History*

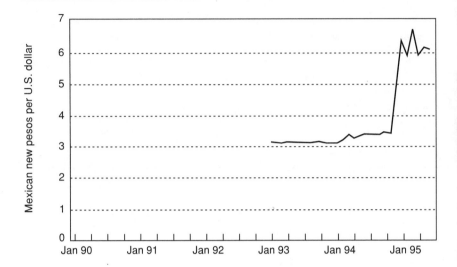

# PURCHASER'S GUIDE TO SINGAPORE

## Major Exports to the United States, 1994

| HS Code | Description | Value, Million Dollars | Growth Rate 94/93 |
|---------|-------------|------------------------|-------------------|
| 8471 | Computers and parts of computers | $5,987 | +13.0% |
| 8473 | Office mach. parts and accessories | 2,533 | +61.3 |
| 8542 | Integrated circuits | 2,001 | +45.0 |
| 2932 | Heterocyclic comp., O-hetero atom | 481 | +3.0 |
| 8527 | Receivers, including phone | 414 | +21.9 |
| 9018 | Medical instruments and parts | 207 | +7.3 |
| 8529 | Parts for radio, TV, phones | 173 | −0.1 |
| 8521 | Video recorders | 166 | +17.6 |
| 8411 | Gas turbines and parts | 141 | +28.0 |
| 2710 | Petroleum not from crude | 137 | −1.7 |
| 6110 | Knitted sweaters, sweatshirts, vests | 121 | −1.8 |
| 8470 | Calculators | 115 | +45.9 |
| 8525 | Transmitters, inc. cordless phone | 112 | −6.5 |

Source: Global Trade Information Services, Inc., Columbia, South Carolina.

## Trade Representation in the United States

| Organization | Phone | Fax |
|--------------|-------|-----|
| Embassy, Washington | (202) 537 3100 | (202) 537 0876 |
| Trade Development Board, New York | (212) 421 2207 | (212) 888 2897 |

## Useful Contacts

| Organization | Phone | Fax |
|--------------|-------|-----|
| American Business Council | +65 235 0077 | +65 732 5917 |
| Arthur Andersen | +65 322 9200 | +65 220 4377 |
| Baker and McKenzie | +65 224 8066 | +65 224 3872 |
| KPMG Peat Marwick | +65 220 7411 | +65 225 0984 |
| Price Waterhouse | +65 225 6066 | +65 225 2366 |

## *Cultural Radar Diagram*

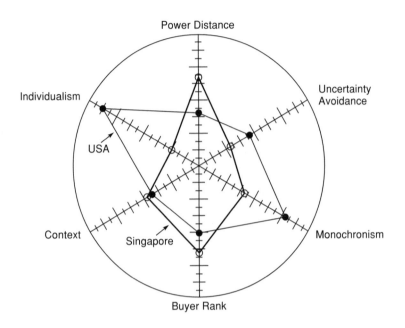

## *Currency History*

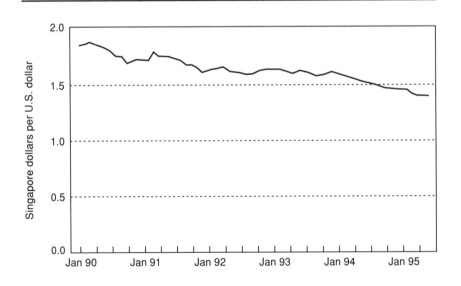

# PURCHASER'S GUIDE TO SOUTH KOREA

## Major Exports to the United States, 1994

| HS Code | Description | Value, Million Dollars | Growth Rate 94/93 |
|---|---|---|---|
| 8542 | Integrated circuits | $3,870 | +59.0% |
| 8703 | Cars | 1,472 | +98.8 |
| 8471 | Computers and parts of computers | 1,294 | −9.8 |
| 8473 | Office mach. parts and accessories | 1,202 | +60.7 |
| 8521 | Video recorders | 612 | −2.5 |
| 8527 | Receivers, including phone | 414 | −12.5 |
| 8516 | Electric heaters, incl. microwave | 414 | +36.0 |
| 6110 | Knitted sweaters, sweatshirts, vests | 412 | +6.9 |
| 8525 | Transmitters, incl. cordless phone | 403 | +40.4 |
| 6403 | Leather shoes | 382 | −32.7 |
| 4203 | Leather clothing | 292 | −43.0 |
| 6204 | Women's suits and ensembles | 276 | −2.8 |
| 6201 | Men's coats, not knitted | 267 | +6.4 |
| 4202 | Luggage | 250 | −4.0 |
| 8523 | Magnetic tape and disks | 235 | −19.3 |
| 7208 | Hot-rolled flat steel | 235 | +87.6 |
| 6205 | Men's shirts | 200 | −11.7 |
| 4011 | Tires | 197 | +12.4 |
| 8518 | Microphones, speakers, amps | 188 | +15.8 |
| 6404 | Shoes with textile uppers | 178 | −27.0 |

Source: Global Trade Information Services, Inc., Columbia, South Carolina.

## Trade Representation in the United States

| Organization | Phone | Fax |
|---|---|---|
| Embassy, Washington | (202) 939 5600 | (202) 797 0595 |
| Consulate, Anchorage | (907) 561 5488 | (907) 563 0313 |
| Consulate, Atlanta | (404) 522 1611 | (404) 521 3169 |
| Consulate, Boston | (617) 348 3660 | (617) 348 3670 |
| Consulate, Chicago | (312) 822 9485 | (312) 822 9849 |
| Consulate, Honolulu | (808) 595 6109 | (808) 595 3046 |
| Consulate, Houston | (713) 961 0186 | (713) 961 3340 |
| Consulate, Los Angeles | (213) 385 9300 | (213) 385 1849 |
| Consulate, Miami | (305) 372 1555 | (305) 371 6559 |
| Consulate, New York | (212) 752 1700 | (212) 888 6320 |
| Consulate, San Francisco | (415) 921 2251 | (415) 931 6330 |
| Consulate, Seattle | (206) 441 1011 | (206) 441 7912 |

## Useful Contacts

| Organization | Phone | Fax |
|---|---|---|
| American Chamber of Commerce | +82 2 753 6471 | +82 2 755 6577 |
| Arthur Andersen | +82 2 767 9114 | +82 2 785 4753 |
| Kim and Kim (attorneys) | +82 2 735 2980 | +82 2 732 3370 |
| Price Waterhouse | +82 2 733 2345 | +82 2 733 5317 |
| U.S. Embassy | +82 2 397 4114 | +82 2 738 8845 |
| KPMG Peat Marwick | +82 2 3442 2345 | +82 2 3443 3200 |

## *Cultural Radar Diagram*

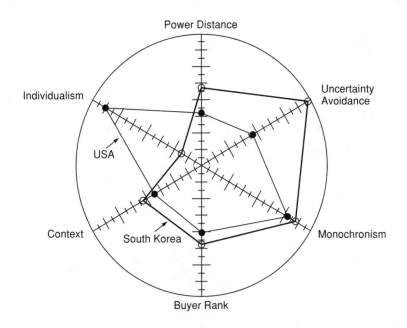

## *Currency History*

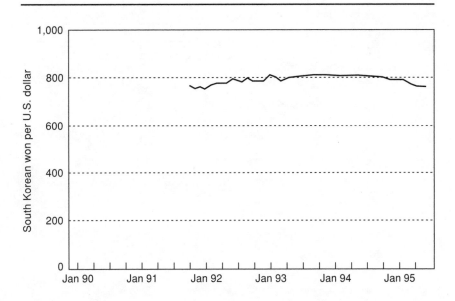

# PURCHASER'S GUIDE TO TAIWAN

## Major Exports to the United States, 1994

| HS Code | Description | Value, Million Dollars | Growth Rate 94/93 |
|---|---|---|---|
| 8471 | Computers and parts of computers | $3,345 | +1.1% |
| 8473 | Office mach. parts and accessories | 2,462 | +38.1 |
| 8542 | Integrated circuits | 1,794 | +41.5 |
| 9403 | Furniture | 780 | −5.2 |
| 9506 | Sports and exercise equipment | 712 | +13.6 |
| 7318 | Nuts, bolts, screws | 619 | +19.6 |
| 9405 | Lamps and lights | 427 | −15.2 |
| 6110 | Knitted sweaters, sweatshirts, vests | 401 | +0.3 |
| 8414 | Air or vacuum pumps | 399 | +13.7 |
| 9401 | Furniture, seats | 395 | −1.4 |
| 8708 | Parts and accessories for vehicles | 345 | +9.9 |
| 8712 | Bicycles and parts | 323 | −7.2 |
| 8534 | Printed circuit boards | 302 | +14.6 |
| 8504 | Transformers | 292 | +4.2 |
| 8518 | Microphones, speakers, amps | 286 | +7.1 |
| 8525 | Transmitters, incl. cordless phone | 274 | +17.7 |
| 6403 | Leather shoes | 270 | −22.8 |
| 8481 | Valves | 262 | +18.8 |
| 8536 | Switches, fuses, breakers | 249 | +12.4 |
| 6204 | Women's suits and ensembles | 247 | −12.1 |

Source: Global Trade Information Services, Inc., Columbia, South Carolina.

## Trade Representation in the United States

| Organization | Phone | Fax |
|---|---|---|
| Taiwan Economic and Cultural Office, Washington | (202) 686 6400 | (202) 363 6294 |
| China External Trade Development Council, New York | (212) 730 4466 | (212) 730 4370 |
| Far East Trade Service, Chicago | (312) 803 8888 | (312) 803 3333 |
| Far East Trade Service, Miami | (305) 477 9696 | (305) 477 9031 |
| Far East Trade Service, San Francisco | (415) 788 4304 | (415) 788 0468 |

## Useful Contacts

| Organization | Phone | Fax |
|---|---|---|
| American Chamber of Commerce | +886 2 581 7089 | +886 2 542 3376 |
| Arthur Andersen | +886 2 545 9988 | +886 2 545 9966 |
| Baker and McKenzie | +886 2 712 6151 | +886 2 716 9250 |
| KPMG Peat Marwick | +886 2 567 6996 | +886 2 567 5802 |
| Price Waterhouse | +886 2 729 6666 | +886 2 757 6371 |

## Cultural Radar Diagram

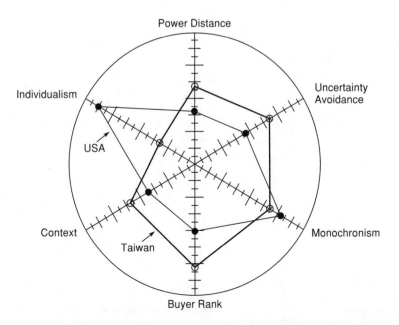

## *Currency History*

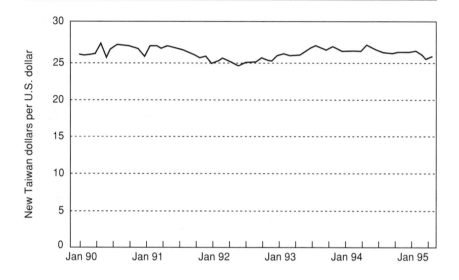

## PURCHASER'S GUIDE TO THAILAND

### Major Exports to the United States, 1994

| HS Code | Description | Value, Million Dollars | Growth Rate 94/93 |
|---|---|---|---|
| 8471 | Computers and parts of computers | $1,408 | +41.8% |
| 0306 | Crustaceans | 835 | +41.0 |
| 8542 | Integrated circuits | 663 | +42.7 |
| 7113 | Jewelry | 363 | +7.0 |
| 8528 | Televisions | 269 | +16.0 |
| 8517 | Telephone & fax equip. and parts | 258 | +0.7 |
| 8521 | Video recorders | 252 | +87.4 |
| 1604 | Processed fish | 237 | +13.7 |
| 6403 | Leather shoes | 236 | +0.8 |
| 4001 | Natural rubber | 201 | +29.1 |
| 8473 | Parts and accessories for computers | 192 | +15.5 |
| 4202 | Luggage and handbags | 191 | +31.1 |
| 8544 | Wire cable and assemblies | 179 | +11.5 |
| 1605 | Processed crustaceans | 165 | +27.2 |
| 7103 | Precious stones, except diamonds | 150 | +18.2 |
| 9009 | Copy machines and parts | 134 | +87.6 |
| 9503 | Toys | 133 | +31.1 |
| 9403 | Furniture | 124 | +0.0 |
| 6110 | Sweaters and pullovers | 120 | −7.4 |
| 2008 | Fruits and nuts | 115 | −16.9 |

Source: Global Trade Information Services, Inc., Columbia, South Carolina.

### Trade Representation in the United States

| Organization | Phone | Fax |
|---|---|---|
| Embassy, Washington | (202) 944 3600 | (202) 429 2949 |
| Thai Trade Center, Atlanta | (404) 659 0178 | (404) 577 6937 |
| Thai Trade Center, Chicago | (312) 467 0044 | (312) 467 1690 |
| Thai Trade Center, Los Angeles | (213) 380 5943 | (213) 380 6476 |
| Thai Trade Center, New York | (212) 466 1777 | (202) 524 0972 |

## *Useful Contacts*

| Organization | Phone | Fax |
|---|---|---|
| American Chamber of Commerce | +66 2 251 9266 | +66 2 255 2454 |
| Arthur Andersen | +66 2 280 0900 | +66 2 280 0855 |
| Baker and McKenzie | +66 2 236 6060 | +66 2 236 6071 |
| KPMG Peat Marwick | +66 2 236 6161 | +66 2 236 6165 |
| Price Waterhouse | +66 2 679 6444 | +66 2 679 6493 |
| U.S. Embassy | +66 2 255 4365 | +66 2 255 2915 |

## *Cultural Radar Diagram*

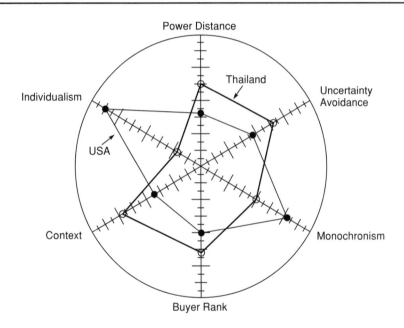

## *Currency History*

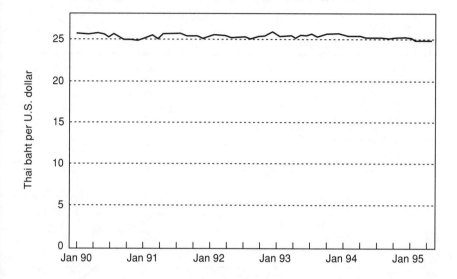

# PURCHASER'S GUIDE TO
# THE UNITED KINGDOM

## Major Exports to the United States, 1994

| HS Code | Description | Value, Million Dollars | Growth Rate 94/93 |
|---|---|---|---|
| 2709 | Petroleum from crude | $2,771 | +31.6% |
| 8411 | Gas turbines and parts | 1,544 | −0.4 |
| 8703 | Cars | 1,022 | +31.5 |
| 8471 | Computers and parts of computers | 826 | +8.1 |
| 2208 | Liquors | 621 | −2.1 |
| 3004 | Antibiotics in doses | 541 | +4.1 |
| 2710 | Petroleum not from crude | 490 | +52.3 |
| 8803 | Parts of aircraft | 473 | −0.6 |
| 8802 | Aircraft and parts | 460 | +53.4 |
| 8473 | Office mach. parts and accessories | 403 | +52.6 |
| 8542 | Integrated circuits | 374 | +9.7 |
| 9701 | Paintings and drawings | 359 | +4.9 |
| 9706 | Antiques | 340 | +14.9 |
| 8708 | Parts and accessories for vehicles | 335 | +36.6 |
| 8701 | Tractors | 333 | +14.9 |
| 2933 | Heterocyclic comp., N-hetero atoms | 304 | +22.6 |
| 7102 | Diamonds | 283 | +1.8 |
| 2922 | Oxygen function amino compounds | 266 | +17.9 |

Source: Global Trade Information Services, Inc., Columbia, South Carolina.

## *Trade Representation in the United States*

| Organization | Phone | Fax |
|---|---|---|
| Embassy, Washington | (202) 462 1340 | (202) 789 6265 |
| Consulate, Atlanta | (404) 524 8823 | (404) 524 3153 |
| Consulate, Boston | (617) 248 9555 | (617) 248 9578 |
| Consulate, Cleveland | (216) 621 7675 | (216) 621 2615 |
| Consulate, Dallas | (214) 637 3600 | (214) 634 9408 |
| Consulate, Houston | (713) 659 6270 | (713) 659 7094 |
| Consulate, Los Angeles | (310) 477 3322 | (310) 575 1450 |
| Consulate, Miami | (305) 374 1522 | (305) 374 8196 |
| Consulate, New York | (212) 745 0495 | (212) 745 0456 |
| Consulate, San Francisco | (415) 981 3030 | (415) 434 2018 |
| Consulate, Seattle | (206) 622 9255 | (206) 622 4728 |

## *Useful Contacts*

| Organization | Phone | Fax |
|---|---|---|
| American Chamber of Commerce | +44 171 493 0381 | +44 171 493 2394 |
| Arthur Andersen | +44 171 438 3000 | +44 171 831 1133 |
| Baker and McKenzie | +44 171 919 1000 | +44 171 919 1999 |
| KPMG Peat Marwick | +44 171 311 1000 | +44 171 311 3311 |
| Price Waterhouse | +44 171 939 3000 | +44 171 378 0647 |
| U.S. Embassy | +44 171 499 9000 | |

## *Cultural Radar Diagram*

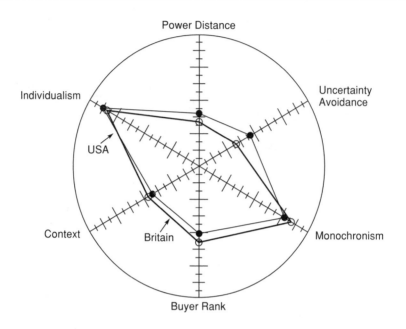

## *Currency History*

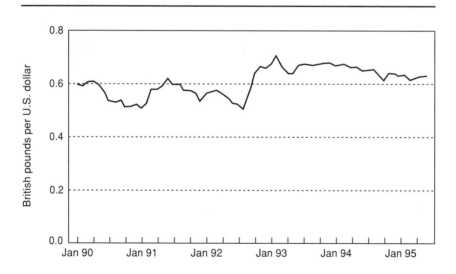

# Index

## A

Action chains, 36
Air freight, 127–130
America, Americans; *see* United States
Andean Pact, 143
Appointments, 36
Arab countries, Arabians, 21, 28, 31
Arbitration, 59
Asia, Asians, 18, 20, 21, 29, 32, 37, 40
Assists, 150–151
Australia, 79

## B

Bowing, 8
Bretton Woods agreement, 67–68
Britain, British, 14, 17–18, 20, 68–69, 286–288
British English language, 45, 49
Brokers, 179
Brokers, customs; *see* Customs brokers
Buyer-seller currency spread, 88

## C

Canada, Canadians, 14, 17, 18, 20, 63, 79, 118, 248–250
Caribbean Basin Initiative, 143
Carriage paid to; *see* Incoterms
Case law, 53
CFR; *see* Incoterms
Change notices, 198
Channels, 173–181
China, Chinese, 53, 60, 118, 251–253
Chinese numbers, 46
Choice of courts, 53, 58
CIF; *see* Incoterms
CIP; *see* Incoterms
CISG, 60–61
Clauses, escape, 121–122
Clusters, 184
Code law, 53

Common law; *see* Case law
Communication, 25–30
  context, 25–29, 39, 40, 245
  formal, 219
  informal, 217
  speed, 25
Communist and formerly communist countries, 56
Conflict, 36
Consensus, 10
Consulates; *see* Embassies and consulates
Containers, ocean freight, 131
Context; *see* Communications
Contracts, 3, 10, 54–61, 211
  attitude toward, 56, 59–60
  complexity, 56
  forward; *see* Forward Contracts
  international, 57–59
Copyrights; *see* Property, intellectual
Cost, landed; *see* Landed cost
Cost and freight; *see* Incoterms
Cost and insurance paid; *see* Incoterms
Cost, insurance and freight; *see* Incoterms
Cost-driven pricing and products, 82–83
Countertrade, 171
Countries, best, 184
Country information, 185
Country of origin marking, 149–150, 210
CPT; *see* Incoterms
Credit terms, 162–163
Cultural advice, 42–43
Cultural difficulties, 3, 5, 7–12
Cultural skills, 9
*Culture's Consequences,* 13–23
*Culturegrams,* 8, 11
Currency
  exchange rate changes, 89
  floating, 79, 82–83
  payment, 78

Currency *(Continued)*
  pegged, 79–81, 82
  pricing, 77–78
  risk, 3
  stronger and weaker, 71
Customs, 137–145, 148–153
  brokers, 129, 145, 153
  costs, 145
  duties; *see* Duties
  export, 128
  U.S. Customs Service, 128, 137

**D**

DAF; *see* Incoterms
DDP; *see* Incoterms
DDU; *see* Incoterms
Decision making, 10
Delivered at frontier; *see* Incoterms
Delivered Duty paid; *see* Incoterms
Delivered Duty unpaid; *see* Incoterms
Delivered ex quay; *see* Incoterms
Delivered ex ship; *see* Incoterms
Delivery, on time, 35
DEQ; *see* Incoterms
DES; *see* Incoterms
Direct purchasing, 176
Documents against payment, 162
Dollar-based pricing, 6, 83
Drafts, 160
Duties, 137–143, 147–149, 150–151
Duty Drawback, 148–149
Duty Reduction Treaties, 141–143

**E**

Embassies and consulates, 2, 189
English language ability, 48
Entertainment, 42
Ethics, 220
EU; *see* European Union
Europe, Europeans, 21, 25, 27, 32–33,
  40, 41, 67, 68, 79
European Union, 68, 118
Ex works; *see* Incoterms
Exchange rates, 69–70
  forward, 70, 93–95
  spot, 86, 87, 93
EXW; *see* Incoterms

**F**

FAS; *see* Incoterms
FCA; *see* Incoterms
Financial data, 192, 195
Flexibility, 197, 210
FOB; *see* Incoterms
Foreign Corrupt Practices Act, 63
Forward contracts, 93–97, 109–110,
  113–114, 116–117
Forwarders, freight; *see* Freight
  forwarders
Forwards; *see* Forward contracts
France, French, 14, 17, 18, 20, 25, 27,
  69, 254–256
Free alongside ship; *see* Incoterms
Free carrier; *see* Incoterms
Free on board; *see* Incoterms
Freight forwarders, 129
Futures, 93, 97–98

**G**

Generalized System of Preferences,
  141–142
Germany, Germans
  culture, 14, 15, 17, 18, 20, 25, 26,
    27, 32, 34, 36, 40–41
  currency, 64, 69, 87
  law, 53, 56
  purchaser's guide, 257–259
Gifts, 36–37
Groups, 19
GSP; *see* Generalized System of
  Preferences

**H**

Hall, Edward and Mildred, 25, 30, 31,
  32–33, 35
Hand gestures, 48
Harmonized Tariff Schedule, 138–140,
  143
Harmony; *see* Values
Hedging, 85–86, 112, 113–114, 116–
  117
Hofstede, Geert, 13, 19, 23
Holidays, 218
Home phones, 218
Hong Kong, 20, 79
Hosting and being hosted, 42

## I

Incoterms, 153–156
India, Indians, 14, 17, 18, 20, 49, 260, 262
Individualism; *see* Values
Indonesia, Indonesians, 2
Insurance, 156–157
Intellectual property, 61–63
Interest rates, 67, 87
Intermediaries, 2, 10, 42, 177
International Procurement Offices, 165–172, 176–177
Interpreters, 5, 48–50
Inventory, 206
Invoice, 144
IPOs; *see* International procurement offices
Israel, Israelis, 14, 17, 143
Italy, Italians, 14, 17, 18, 20, 34, 68–69, 263–265

## J

Japan, Japanese
  culture, 9, 14–15, 17, 18, 19, 20, 31, 32, 34, 36
  communication, 27, 28, 47, 48–49
  companies, 1
  currency, 79
  economy, 41
  language, 29
  law, 53, 64
  purchaser's guide, 266–268
JIT, 134

## K

Korea, Koreans; *see* South Korea, South Koreans

## L

Labor stability, 196
Landed cost, 203–207
Language difficulties, 3, 45–52
Latin America, Latin Americans, 30, 31
Law, legal, 3, 10, 53–65
Lead time, 35, 197
Letters of credit, 159–161
Licenses, 152
Logistics, 127–135

## M

Macroeconomics, 2
Magazines, 185–188
Malaysia, Malaysians, 58, 269–271
Manners, 8
Manufacturing processes, 196–197
Market-driven pricing and products, 81, 83
Masculinity; *see* Values
Materials management, 197
Measurement, supplier, 214–215
Meetings, documenting, 50
Mexico, Mexicans, 14, 17, 18, 20, 33, 34, 78, 79, 80, 272–274
Monochronic, monochronism; *see* Time
Motivation, 10
Multiple currencies, 118–119

## N

NAFTA, 142–143
National Trade Data Bank, 186–187
Negotiations, final, 209–210
New Zealand, 79
Nigeria, Nigerians, 2
No, getting to, 29–30

## O

Ocean freight, 131–133
Options, 101–107, 116–117

## P

Patents; *see* Property, intellectual
Pegged currency; *see* Currency
Poland, 118
Political stability, 187
Polychronic, polychronism; *see* Time
Porter, Michael, 184
Power distance; *see* Values
Privacy; *see* Space, territory and privacy
Process control, statistical; *see* Quality
Property, intellectual, 61–63

## Q

Quality
  AQL, 199
  inspection, 210

Quality *(Continued)*
  ISO 9000 series, 198–199
  statistical process control SPC, 5, 16, 196
Questions, negative, 47
Quotas, 150

**R**

Rates, exchange; *see* Exchange rates
Rates, spot; *see* Exchange rates
Reciprocity, 63
References, 192
Representatives, buyers, 43, 176–177
Representatives, sellers; *see* Intermediaries
Risk, 207–208
  air freight, 207
  exchange, 71–73
  political, 187
Risk sharing, 122–124

**S**

Singapore, Singaporeans, 14, 17, 18, 20, 49, 79, 275–276
South Korea, South Koreans, 14, 17, 18, 20, 78, 277–279
Space, territory and privacy, 31–32
Speculation, currency, 85
Statistical process control; *see* Quality
Stereotyping, 9, 40
Subsidiaries, sellers', 173–175
Suppliers
  eliminating potential, 191–209
  feedback, 215
  managing, 198–213
  measuring, 214–215
  reviews, 219
  selecting, 183–201
  size, 195
  surveying, 193–200
Sweden, Swedes, 20

**T**

Taboos, 8
Taiwan, Taiwanese, 20, 41, 79, 80–81, 118–119

Terms of sale, 4; *see also* Incoterms
Thailand, Thais, 14, 17, 18, 20, 283–285
Time monochronic and polychronic, 32–36, 39, 40, 246
Time, 32–36
Title transfer; *see* Incoterms
Tooling, duties on, 6, 150–151
Trade Offices, 188–189
Trademarks; *see* Property, intellectual

**U**

U.S. Chamber of Commerce, 49, 189
U.S. goods returned, 147–148
Uncertainty avoidance; *see* Values
*Understanding Cultural Differences,* 25, 30
United States, Americans, 9, 14, 17, 18, 19, 20, 21, 27, 28–29, 31, 33, 40, 41–42, 47, 63
United Kingdom, British; *see* Britain, British

**V**

Values
  buyer rank, 15, 39, 246
  core purchasing, 214
  face, 21
  guilt, 21
  harmony, drive toward, 21–22
  individualism, 18, 39, 40, 245
  masculinity, 19–20
  power distance, 13–16, 39, 40, 244–245
  shame, 21
  uncertainty avoidance, 16–17, 39, 40, 245
Visiting suppliers, 210
Volumetric weight, 130

**W**

*Wall Street Journal, The,* 69
Warranty, 210
Wire transfer, 162
Women in international purchasing, 21

# About the Author

DICK LOCKE, before founding the Global Procurement Group in January of 1993, spent 15 years in Hewlett-Packard Procurement. He managed HP's corporate semiconductor procurement engineering section and led HP's first aggressive approach to improving purchased semiconductor quality. He then managed HP's first multidiscipline procurement team for semiconductors. This group made major improvements in assurance of supply, cost, flexibility, and quality.

The last eight years were in management positions in the International Procurement group. While living in Tokyo, he founded and managed HP's International Procurement Operations in four Asian countries. Later, he managed offices in Europe and Mexico.

He is a member of the National Association of Purchasing Management and has written for their publications. He also is on the "Purchasing Issues" advisory board for *Electronic Buyer's News.*

His education includes a BSEE with a concentration in computer science from Stanford University, and an MBA from Santa Clara University.

Through his company, Global Procurement Group, he presents seminars and consults on purchasing-related topics. The company can be reached by phone at (415) 695-1673, or by e-mail at info@globalpg.com.

Other books of interest to you from Irwin Professional Publishing . . .

## THE ISO 9000 HANDBOOK, SECOND EDITION

### Edited by Robert W. Peach

This is an extraordinarily detailed, timely, and easy-to-use resource which
provides an in-depth, clause-by-clause explanation of the Q9000 series text.
Walks the reader through all phases of implementation, and examines the
broader issues of product liability, conformity, assessment, and registrar
accreditation.
ISBN: 0-883337-31-3        700 pages

## ISO 14000

### A Guide to the New Environmental Management Standards

### Tom Tibor

A comprehensive, invaluable preview of the ISO standards for companies that
want to know what ISO 14000 is, how it will affect them, and what they can do
about it. Includes a complete description of each standard, and discusses the
implications each standard will have for international trade and regulatory
enforcement.
ISBN: 0-7863-0523-1        150 pages

## QUALITY SYSTEMS UPDATE

### A Global ISO 9000 and ISO 14000 Information Service

Subscribe to *QSU* and you'll receive 12 issues of the monthly newsletter that
has been acclaimed by thousands of quality professionals as the preeminent
publication on ISO 9000 quality systems and related standards development. A
"hard news" newsletter, *QSU* is the industry's leading news source—and an
important management tool. Join the professionals who rely on *QSU* to stay
informed. To order, please call (703) 591-9008.
ISBN: 1060-1821        $395.00 per year